Mathematical Nuggets from Austria

**Selected Problems from
the Styrian Mid-Secondary School
Mathematics Competitions**

Problem Solving in Mathematics and Beyond

Print ISSN: 2591-7234
Online ISSN: 2591-7242

Series Editor: Dr. Alfred S. Posamentier
Distinguished Lecturer
New York City College of Technology - City University of New York

There are countless applications that would be considered problem solving in mathematics and beyond. One could even argue that most of mathematics in one way or another involves solving problems. However, this series is intended to be of interest to the general audience with the sole purpose of demonstrating the power and beauty of mathematics through clever problem-solving experiences.

Each of the books will be aimed at the general audience, which implies that the writing level will be such that it will not engulfed in technical language — rather the language will be simple everyday language so that the focus can remain on the content and not be distracted by unnecessarily sophiscated language. Again, the primary purpose of this series is to approach the topic of mathematics problem-solving in a most appealing and attractive way in order to win more of the general public to appreciate his most important subject rather than to fear it. At the same time we expect that professionals in the scientific community will also find these books attractive, as they will provide many entertaining surprises for the unsuspecting reader.

Published

For the complete list of volumes in this series, please visit www.worldscientific.com/series/psmb

Problem Solving in
Mathematics and Beyond | Volume 19

Mathematical Nuggets from Austria

Selected Problems from the Styrian Mid-Secondary School Mathematics Competitions

Robert Geretschläger

BRG Keplerstraße, Austria

Gottfried Perz

BG/BRG Pestalozzistraße, Austria

World Scientific

NEW JERSEY · LONDON · SINGAPORE · BEIJING · SHANGHAI · HONG KONG · TAIPEI · CHENNAI · TOKYO

Published by

World Scientific Publishing Co. Pte. Ltd.
5 Toh Tuck Link, Singapore 596224
USA office: 27 Warren Street, Suite 401-402, Hackensack, NJ 07601
UK office: 57 Shelton Street, Covent Garden, London WC2H 9HE

Library of Congress Cataloging-in-Publication Data
Names: Geretschläger, Robert, author. | Perz, Gottfried, author.
Title: Mathematical nuggets from Austria : selected problems from the
    Styrian mid-secondary school mathematics competitions / Robert Geretschläger,
    BRG Keplerstraβe, Austria, Gottfried Perz, BG/BRG Pestalozzistraβe, Austria.
Description: New Jersey : World Scientific, [2020] |
    Series: Problem solving in mathematics and beyond, 2591-7234 ; vol. 19
Identifiers: LCCN 2020020784 | ISBN 9789811219894 (hardcover) |
    ISBN 9789811219252 (paperback) | ISBN 9789811219269 (ebook) |
    ISBN 9789811219276 (ebook other)
Subjects: LCSH: Mathematics--Problems, exercises, etc. | Mathematics--Competitions--
    Austria--Styria. | Mathematics--Study and teaching (Middle School)--Austria--Styria.
Classification: LCC QA43 .G43 2020 | DDC 372.7/044--dc23
LC record available at https://lccn.loc.gov/2020020784

British Library Cataloguing-in-Publication Data
A catalogue record for this book is available from the British Library.

For any available supplementary material, please visit
https://www.worldscientific.com/worldscibooks/10.1142/11796#t=suppl

Desk Editors: V. Vishnu Mohan/Tan Rok Ting

Typeset by Stallion Press
Email: enquiries@stallionpress.com

Printed in Singapore

Preface

We hope that you will enjoy this collection of mathematical puzzles. The problems in this book were developed specifically for problem-solving competitions for students aged 11 to 14, but they are meant to pose enough of a challenge to make them interesting for problem-solvers of all ages.

The questions posed here are sometimes stated in a very mathematical way and sometimes cased in the shell of a real-world situation. However, the kernel of each of these puzzles is always quite abstract. The goal in each case is to consume the information given in the respective problem setting, digest it all through application of pure logic and common sense, perhaps spiced up with a little bit of elementary calculation on the side, and then use the energy resulting from the resolution of the problem to develop our intellectual muscles one little piece at a time. This is meant as a purely enjoyable exercise, even if there can be some frustrating moments in the process. The satisfaction we feel after solving such problems is, of course, derived from overcoming these very intellectual obstacles. If there were no mental resistance to defeat, there would be no feeling of accomplishment.

The authors would like to give special thanks to V. Vishnu Mohan and Tan Rok Ting for their roles in the production of this book, Alfred Posamentier for his part as series editor, and to our spouses, Zita and Sylvia, for their constant loving support and encouragement.

About the Authors

Robert Geretschläger is a mathematics teacher at Bundesrealgymnasium Keplerstraße in Graz, Austria, and a lecturer at the Karl-Franzens University in Graz. He is involved in the organization of several mathematics competitions, among other things as a leader of the Austrian team at the International Mathematical Olympiad and a main organizer of the Mathematical Kangaroo in Austria. He is the editor of the two-volume *Engaging Young Students in Mathematics through Competitions* and co-author of *A Central European Olympiad*, among other works.

Gottfried Perz is a teacher of mathematics and geometry at Pestalozzi Gymnasium in Graz, Austria. As a former participant in the Austrian Mathematical Olympiad, he was involved in the organization of various mathematical competitions throughout his professional life. For several years, he was the main organizer of the Steirische Unterstufenwettbewerb.

Contents

Introduction

As we all know, math is a lot of fun. Well, ok, not everyone knows this, but an awful lot of us do, and anyone reading this is likely to agree with the idea that math is, indeed, a lot of fun.

One aspect of math that a lot of people find really enjoyable is solving logical puzzles. For some people, this means looking forward to the Sudoku in their daily newspaper, assuming that they still read newspapers. For others, it means working out a challenge posed by a fire-breathing dragon in a medieval-looking computer game to gain entrance to a well-guarded castle and the enchanted princess within. There are many different ways to package logical posers, but finding that Aha moment when we finally comprehend the complexities of a problem we have been struggling with is something many of us can appreciate, no matter what container the task may be wrapped up in.

One possible way to get those students who might not yet appreciate mathematics interested in the subject is to show them how much fun it can be to solve such problems, and one possible avenue for this is the world of Mathematics Competitions. Problems, as they are typically set at mathematics competitions all around the world, are usually somewhat close to the math being taught in the classroom, but not quite the same. The tools that students acquire in a regular math class may prove to be useful in solving a competition problem, but the most important thing will always be their own original idea. The insight into the structure of the situation will always be that of the person who is solving it, and discovering that covert information in a flash of logical realization is what the joy of problem solving is all about.

The problems in this book were all posed at math competitions for grade levels 6–8 in the last quarter century in Styria (Steiermark, in German), one of the nine provinces in Austria. If you are not so familiar with this part of the world, you will find Styria in the south-east of Austria on the map of Europe, just north of the border to Slovenia. The problems have been taken from three competitions. The one with the widest reach is the *Steirische Unterstufenwettbewerb* (StU for short). This is an annual competition that started in 1989 and is still going strong. Students from schools all over the province are invited to a central spot (BRG Körösi in the capital city, Graz) on a Wednesday in April to solve (since 1996) ten multiple-choice problems and seven full-answer questions. Another competition with a tradition going back almost as far is the so-called *Schul5Kampf* (or S5K). The participating schools have varied in number over the years from three to five, but the structure has been the same throughout, with participants solving ten problems, the answer to each of which is just a number. This team competition, held at the end of each school year, is generally considered by the people involved to be one of the most enjoyable aspects of the last weeks of school, just before people head off for their summer vacations. Finally, there is the *OMMO*, or *Obermurtal Mathematikolympiade*. This was a short-lived competition that only managed to survive for a few years, but the structure was similar to the S5K, in that it involved schools from a certain area, namely the northern part of Styria. The problems posed at this competition were all of full-answer style.

The great majority of the problems collected in this book were posed by the authors, who have both been active in setting competition problems for decades. Some were suggested by other local mathematicians and teachers. Since records of authors who provided the problems were not maintained over the years, any listing of them will necessarily be incomplete, but important contributions came from Gerhard Plattner, Josef Prechtler and Erich Windischbacher, among others. We thank all those who have suggested problems for these competitions over the years.

This book is divided into sections by topic. Some topics are of a mathematical nature (like circles or sequences), while others are descriptive in nature (like money or sports). Therefore you will find a few problems in two different sections. You will find that the level of difficulty varies quite a bit from problem to problem. Each section is made up of three parts. In the first part, unsurprisingly headed Problems, the questions are asked, and a brief footnote is given with reference to the competition the problem was posed at. (If, for instance, the footnote reads (StU-08-B2), this problem was number B2, posed in 2008 at the StU competition. StU refers to Steirische Unterstufenwettbewerb, S5K refers to the Schul5Kampf, and O refers to OMMO.) You are encouraged to try to solve each problem as it tickles your curiosity. Solving the problems in the order they are presented may be helpful in some cases, but is not important. The second part, headed Tips, gives you a hint toward solving the problem if you can't quite come up with a solution yourself, without giving it all away, and leaving something left to be discovered. Sometimes such hints can get you started, and the satisfaction of finding the solution is then still to be had. Finally, if you either can't find a solution or would like to compare your ideas to the ideas of the authors, there is a section headed Solutions. Here, you can find complete arguments for each of the answers. Sometimes, you might find it useful to read through one or two of these, before trying your hand at the next problem.

Over the years, many of the participants in these competitions have advanced to higher level ones, like the regional and national Olympiads, or even to international competitions. This is, however, not the main focus here. If we can get as many students as possible to appreciate the intellectual pleasure to be derived from abstract thoughts, we have done our job. Most participants in these competitions will not go on to pursue a career in mathematics, but with any luck, all participants will remember the joy that was derived from applying themselves to arrive at a solution to the problems. We hope that you, the reader, will experience some of this joy in working through the problems as well!

Chapter 1

Number Digits

Sometimes, the most fun you can have with numbers comes from just playing around with their digits. There are all kinds of weird and enjoyable things that turn up when you start thinking about combinations of digits that numbers with certain properties can have. This chapter contains some problems of this type. You may be confronted with numbers whose digits have interesting sums or products, or whose order might change in interesting ways under certain circumstances. The most important thing is to keep an open mind; you'll be surprised at what you will be able to figure out. And remember, if you can't get the solution right away, you can find a tip to help you on your way at the end of the problem list.

PROBLEMS

Problem 1.1:
How many six-digit positive integers have 2 as the sum of their digits?

(StU-17-A9)

Problem 1.2:
How many four-digit positive integers have 9 as the product of their digits?

(StU-12-A2)

Problem 1.3:
How many positive integers less than 100 have 6 as the sum of their digits?

(O-93-4)

1

Problem 1.4:
How many positive integers smaller than 1000 have digits that sum to 6?

(O-93-4*)

Problem 1.5:
How many three-digit numbers exist, in which the ones-digit is equal to the sum of the hundreds-digit and the tens-digit?

(StU-99-A6)

Problem 1.6:
How many two-digit positive integers are seven times as large as the sum of their digits?

(StU-09-A9)

Problem 1.7:

(a) Determine the smallest positive integer, the sum of whose digits is equal to 97.
(b) Determine the smallest positive integer, the product of whose digits is equal to 29 400.

(StU-97-B7)

Problem 1.8:
Determine a positive integer with the property that the product of its digits equals 2160. Why can such an integer not have only three digits? What is the smallest integer with this property?

(StU-03-B2)

Problem 1.9:
What is the largest possible remainder that can remain after dividing a two-digit number by the sum of its digits?

(O-94-5)

Problem 1.10:
Determine both the largest and smallest possible numbers that can be attained by adding a two-digit number to the sum of its own digits.

(O-95-4)

Problem 1.11:
Determine all four-digit numbers n with the property that the sum of n and the sum of the digits of n is equal to 2016.

(StU-16-B1)

Problem 1.12:
The digits of a three-digit number N are all different. The product of the digits of N is divisible by 27, but not by any prime other than three. Determine the sum of the digits of N.

(StU-14-B7)

Problem 1.13:
Two different numbers are called "*in-laws*" if each is divisible by the sum of the digits of the other. How many two-digit in-laws does the number 24 have?

(O-01-3)

Problem 1.14:
We are given two sets of three-digit numbers. The set A contains all three-digit numbers, the sum of whose digits is equal to 6. The set B contains all three-digit numbers in which the ones-digit is smaller than the tens-digit by 1, and the hundreds-digit is larger than the tens-digit by 1. Is it possible to choose a number a from set A and a number b from set B, such that the greatest common divisor of a and b is equal to one?

(StU-98-B5)

Problem 1.15:
A number like 424 or 167761 that reads the same from left to right and from right to left is called a palindrome. How many even four-digit numbers are palindromes?

(StU-02-A4)

Problem 1.16:
Just as Mr. Leadfoot is getting ready to drive off in his hot-rod, he notices that his odometer reads 37654 km. He notes that this is a number made up of five consecutive digits, namely 3, 4, 5, 6 and 7. How many kilometers does

Mr. Leadfoot have to drive until his odometer once again reads a number composed of five consecutive digits?

(O-03-1)

Problem 1.17:
Caroline multiplies a three-digit number N by 9. Amazingly, as a result she gets a four-digit number in which the digits, read from left to right, are four consecutive increasing numbers. Does such a number actually exist? If so, what is the number N? If not, why not?

(StU-06-A9)

Problem 1.18:
Consider all positive integers divisible by 8 with ascending digits, i.e. numbers (like, for instance, 136) with the property that the tens-digit is larger than the hundreds-digit, and the ones-digit is larger than the tens-digit. What is the hundreds-digit of the largest of these numbers?

(StU-11-A3)

Problem 1.19:
How many numbers are there from 1 to 1000 that are written without either the digit 0 or the digit 9?

(StU-92-8)

Problem 1.20:
In the far-off island nation of Trenixa, the people are incredibly afraid of the number three. When they count, they not only leave out all numbers divisible by three, but also all numbers that are written with the digit 3 (like 23 or 134, for instance). This year, the Trenixans will be celebrating the "100"th birthday of their Grand Whoopty, according to their counting method. How old will the Grand Whoopty really turn this year?

(O-99-2)

Problem 1.21:
A is the biggest four-digit number that has exactly three different digits. B is the smallest four-digit number that has exactly three different digits. Determine the number $A - B$.

(StU-16-A5)

Problem 1.22:
The year 1998 has a number with the property that it is twice a number (999) that is written by repetition of a single digit. When was this last the case before 1998? When will it next be the case again after 1998? How many years will have passed between those two years?

(O-98-1)

Problem 1.23:
How many three-digit positive integers are there, that are written using exactly two different digits?

(StU-97-B6)

Problem 1.24:
The product of two two-digit numbers is a number written only using the digit 8 repeatedly. How many digits can this number have at most and how many must it have at least? Determine all possible combinations of pairs of two-digit numbers yielding such a product.

(StU-00-B4)

Problem 1.25:
How many positive integers smaller than 200 are there with a decimal representation that contains exactly two different digits?

(StU-08-A9)

Problem 1.26:
We consider the numbers $a = 999\ldots9$ (composed of 20 times the digits 9) and $b = 999\ldots9$ (composed of 12 times the digit 9). Determine the sum of the digits of $a \times b$.

(StU-14-B5)

Problem 1.27:
In a seven-digit number N divisible by 6, the first, third, fifth and seventh digit are equal, and the second, fourth and sixth digit are equal (i.e. the decimal representation of N is *ababab*). If you delete the first and the last digit, you get a five-digit number (with decimal representation *babab*), that

is divisible by 9, but not divisible by 18. What is the second digit b of the given number N?

(StU-08-A10)

Problem 1.28:
Determine the smallest positive integer with the following properties: The first digit of the number is 6. If we remove this first digit and tack it onto the end of the number, we obtain a new integer, whose value is exactly one-quarter that of the original number.

(StU-90-7)

Problem 1.29:
In the ten-digit number 4676584251, five digits are to be deleted in such a way, that the remaining five digits form a five-digit number, which is divisible by 5 and as large as possible. What is the sum of the deleted digits?

(StU-08-A6)

Problem 1.30:
A five-digit number divisible both by 5 and by 9 has the following properties. Its hundreds-digit is equal to 6, and its first and last digits are the same. If you delete the first and the last digit, you obtain a number divisible by 4. Determine all such numbers and give reasons why there cannot be more numbers with these properties.

(StU-07-B4)

Problem 1.31:
Two two-digit numbers are formed from the digits 8, 7, 5 and 2. What is the largest possible value of the product of two numbers obtained this way?

(StU-00-A3)

Problem 1.32:
Charlie does not like to write identical digits side-by-side when he writes numbers. In order to avoid this, he has developed an alternative writing system that counts the number of neighboring identical digits in a number. Instead of writing the number 333788888, he counts the threes, sevens and eights and writes the expression 3_37_18_5. When he writes 6_27_11_5, he

is using this expression to denote the number 66711111. Which expression would he write as the result of the addition 4_15_36_2 + 4_56_1?

(StU-06-A6)

Problem 1.33:
Caroline writes a five-digit number N divisible by 4 using each of the digits 1, 2, 3, 4 and 5 once. Which of the digits cannot be the tens-digit of this number?

(StU-07-A5)

Problem 1.34:
The number *googol* is defined as 10^{100}. How many nines are there in the number we obtain by subtracting 1994001994 from a googol?

(O-94-4)

Problem 1.35:
The (four-digit) number of a year can be turned upside-down if the number is made up only of digits 0, 1, 6, 8 and 9. In the year 1996, this was possible, for instance, and calculating the difference between the number of the year and its upside-down version gives us $9661 - 1996 = 7665$. Determine the largest and smallest possible non-negative differences we can obtain by determining such differences. (Note that the first digit of the number of a year cannot be 0.)

(O-96-2)

TIPS

Tip for Problem 1.1:
If the sum of six digits is less than 6, some of those digits will have to be zeros.

Tip for Problem 1.2:
Find out which digits the decimal representation of a four-digit number with this property might consist of.

Tip for Problem 1.3:
How many digits can numbers less than 100 have?

Tip for Problem 1.4:
If a number is smaller than 1000, it can have 1, 2 or 3 digits. These three cases will need to be considered separately when we are thinking about the sums of the digits.

Tip for Problem 1.5:
Consider the answer to this if you know the specific value of the ones-digit. How many such numbers are there with the ones-digit 0? How many with the ones-digit 1? How about 2?

Tip for Problem 1.6:
If a number is seven times as large as another number, that means it must be divisible by seven.

Tip for Problem 1.7:
If we want a number to be as small as possible, this means that it should have as few digits as possible, and that its lead digit should be as small as possible.

Tip for Problem 1.8:
This is very similar to part (b) of the last problem.

Tip for Problem 1.9:
First of all, think about what the largest possible value of the sum of the digits of a two-digit number is. The rest remaining after any division is always smaller than the divisor.

Tip for Problem 1.10:
This one is easy. If a number has a fixed number of digits, it will be big if the digits are big and small if the digits are small.

Tip for Problem 1.11:
This may seem to be almost the same as the last problem, but there is much more to this one. How many digits must a number with this property have? Can it have three? What could the first digit of such a number be?

Tip for Problem 1.12:
What do we know about each individual digit of the number under the given circumstances?

Tip for Problem 1.13:

Take a look at the sum of the digits of 24. This number must be a divisor of all its in-laws. Also, take a look at the divisors of 24. These are candidates for the sum of the digits of its in-laws. That doesn't leave a whole lot of options, does it?

Tip for Problem 1.14:

You might have to think a bit sideways to get this one. What do you know about a number if you know that the sum of its digits is equal to 6?

Tip for Problem 1.15:

If you know the ones-digit of a four-digit palindrome, you also know its thousands-digit. Similarly, if you know its tens-digit, you also know its hundreds-digit.

Tip for Problem 1.16:

Can the first digit of the next such number be 3 again? What would the second digit be in that case?

Tip for Problem 1.17:

We have much more information about the result of Caroline's multiplication than about N. It would therefore make sense to work backwards.

Tip for Problem 1.18:

Any natural number divisible by 8 is certainly even, and the ones-digit of a number of this type can therefore not be 9. This tells us something about the tens-digit, as well. Furthermore, the number in question must obviously by divisible by 4.

Tip for Problem 1.19:

It will be a good idea to consider one- two- and three-digit numbers separately.

Tip for Problem 1.20:

How many numbers from 1 to 100 are divisible by 3? How many are written using the digit 3? How many numbers are there with both properties?

Tip for Problem 1.21:

A big number with a specified number of digits has a big first digit. Similarly, a small number with a specific number of digits has a small first digit.

Tip for Problem 1.22:
9 is the biggest digit, of course. If we want a bigger number with the property we are thinking about, we will not just be able to jump to the next digit, because there isn't one. What else could we do to get a bigger number with all digits the same?

Tip for Problem 1.23:
Think about the number of different ways you can write a three-digit number with exactly two different digits. There aren't all that many, right? Also, where can 0 go in a three-digit number (or rather, where can't it go)?

Tip for Problem 1.24:
How many digits do the products 10×10 and 100×100 have?

Tip for Problem 1.25:
The numbers in question can either be two-digit numbers with two different digits, or three-digit numbers containing one digit twice and the other digit once. Since these numbers must be smaller than 200, there is only one option for their hundreds-digit.

Tip for Problem 1.26:
Both of the given numbers are one less that a power of ten!

Tip for Problem 1.27:
You should know how to recognize a number divisible by 9. Also, if a number is divisible by 9 but not by 18, you know that that number must be odd.

Tip for Problem 1.28:
Take another look at the tip for the last problem. This idea will help you again here, even if there is an additional wrinkle to this problem.

Tip for Problem 1.29:
What does the last digit of the resulting number have to be? What does the first digit of the resulting number therefore have to be?

Tip for Problem 1.30:
Note that the first digit of a five-digit number cannot be equal to 0. What does this tell us about the last digit, if we know that the number is divisible by 5?

Tip for Problem 1.31:
Which of the given digits must be the tens-digits of the two numbers? Once you have decided on those, there aren't really that many cases to check.

Tip for Problem 1.32:
This comes down to interpreting Charlie's notation. Note that 3_3 means 333, 7_1 means 7 and 8_5 means 88888.

Tip for Problem 1.33:
Obviously, the ones-digit of N must be even.

Tip for Problem 1.34:
Try subtracting 1 from a googol first. What special property does the result have?

Tip for Problem 1.35:
To obtain a big difference, you will have to subtract something very small from something very large. Recall that large numbers have large initial digits. To obtain a small difference, you will want to get the lead digits as close together as possible. Is it possible for the difference to have less than four digits?

SOLUTIONS

Solution for Problem 1.1:
There are two ways in which six digits can add up to 2. Either four of the digits are zeros and two are ones, or five are zeros and one is a two. In the second case, there can only be one such number, as the two must be the first digit, since the first digit of a number can never be a zero. In the first case, the first digit must be one of the ones for the same reason, but the other one can be any one of the remaining five digits, and there are therefore five such numbers. We see that there are six numbers of this type altogether, namely

200000, 100001, 100010, 100100, 101000 and 110000.

Solution for Problem 1.2:
The only digits dividing 9 are 1, 3 and 9. The four-digit numbers in question must therefore not contain any digit other than these in their decimal representations. This means that we have to determine the number of four-digit

natural numbers that consist of either two digits 1 and two digits 3 or of three digits 1 and one digit 9.

In the first case, we get the numbers 1133, 1313, 1331, 3113, 3131 and 3311, and in the second case, we get the numbers 1119, 1191, 1911 and 9111.

In total, there are therefore ten four-digit numbers, whose digits have 9 as their product.

Solution for Problem 1.3:

A positive integer less than 100 can have either one or two digits. The only one-digit number with 6 as the "sum" of its digits is 6 itself. All we therefore really need to do is count the number of two-digit numbers with this property. The first digit (i.e. the tens-digit) must be a number from 1 to 6, and the ones-digit is uniquely determined by the tens-digit, since they must sum to 6. The only possible numbers of this type are therefore

$$15, 24, 33, 42, 51 \text{ and } 60.$$

Together with the single-digit number 6, this gives us 7 numbers with the required property.

Solution for Problem 1.4:

First of all, we can take a look at the numbers with only one digit. For these, the single digit is already the "sum" of the digits, and the only such number is the number 6. This case is the easiest to deal with.

Now we can turn to the two-digit numbers. This case is a bit more difficult, but not much. The tens-digit of such a number cannot be greater than 6, because the sum of the digits would then definitely be greater than 6. It can, however, be any of the digits from 1 to 6, since the ones-digit is then simply whatever is missing to the sum 6. (If the tens-digit is 4, for instance, the ones-digit must be $6 - 4 = 2$, and the number is therefore 42.) There must therefore be 6 such two-digit numbers, namely 15, 24, 33, 42, 51 and 60.

This idea will also help us deal with the most difficult case, namely that of the three-digit numbers. The hundreds-digit can be 1, 2, 3, 4, 5 or 6. In each of these cases, there will exist as many three-digit numbers with this hundreds-digit as there are two-digit numbers with whatever is missing to the sum 6 as the sum of their digits. Note that these two-digit numbers can also begin with 0, of course. What does this mean? Let us assume, for example, that 4 is to be the hundreds-digit. There are as many three-digit numbers with the sum of their digits being 6 and their hundreds-digit 4, as there are two-digit numbers with the sum of their digits being $6 - 4 = 2$. There are three such numbers, namely 02, 11 and 20. The corresponding three-digit numbers are 402, 411 and 420.

How many such numbers are there then in total? With hundreds-digit 6, there exist as many as there are two-digit numbers with the sum of their digits being $6 - 6 = 0$, namely 1 (specifically, 00). Similarly, with hundreds-digit 5, there exist as many as there are two-digit numbers with the sum of their digits being $6 - 5 = 1$, namely 2. With hundreds-digit 4, there exist 3, with hundreds-digit 3, there exist 4, with hundreds-digit 2, there exist 5 and with hundreds-digit 1, there exist 6. Altogether, this gives us $1 + 2 + 3 + 4 + 5 + 6 = 21$ three-digit numbers with the sum of their digits being 6.

These numbers can also be written out explicitly:

600					
501	510				
402	411	420			
303	312	321	330		
204	213	222	231	240	
105	114	123	132	141	150

How many numbers with the sum of their digits being 6 are there then altogether? We know that there are 1-, 2- and 3-digit numbers of this type. Adding them up, we see that there exist

$$1 + 6 + 21 = 28$$

such numbers.

There is another way to do this as well. We can think of numbers with less than three digits as numbers whose hundreds-digit (and possibly also their tens-digit) is 0. If we think of the problem in this way, we do not need to differentiate the one- and two-digit numbers from the three-digit ones, since we can simply say that their total number is equal to the number of two-digit numbers with the sum of their digits being 6 (with 0 allowed as a tens-digit as well). There are 7 such numbers, and, with the same reasoning we used before, the total number of three-digit numbers with the sum of their digits being 6 (and 0 allowed as the hundreds-digit) is therefore equal to

$$1 + 2 + 3 + 4 + 5 + 6 + 7 = 28.$$

Solution for Problem 1.5:
If the ones-digit of a three-digit number N is equal to the sum of its hundreds-digit and its tens-digit, neither the hundreds-digit nor the tens-digit of N can be larger than its ones-digit. Furthermore, the hundreds-digit of a three-digit number cannot be 0. This means that there are no such numbers N with the ones-digit 0.

We can now consider what happens whenever we raise the value of the ones-digit of N by 1. If the ones-digit of N is 1, the hundreds-digit must also be 1, and the tens-digit must therefore be $1 - 1 = 0$. The only such number is therefore 101.

If the ones-digit of N is 2, the hundreds-digit h of N can be either 1 or 2, with the tens-digit being equal to $2 - h$. There are two such numbers, namely 202 and 112.

If the ones-digit of N is 3, the hundreds-digit h of N can be either 1, 2 or 3, with the tens-digit being equal to $3 - h$. There are three such numbers, namely 303, 213 and 123.

A similar idea shows us what happens for any ones-digit n. For any given n and any hundreds-digit h from 1 to n, the tens-digit must be equal to $n - h$, yielding n numbers with the required property. This means that the total number of three-digit numbers with the required property is equal to

$$1 + 2 + 3 + 4 + 5 + 6 + 7 + 8 + 9 = 45.$$

Solution for Problem 1.6:
If we know that a number is equal to seven times another number, it must be a multiple of seven. There are not that many two-digit multiples of seven, and it is not difficult to simply make a complete list of them:

$$14, 21, 28, 35, 42, 49, 56, 63, 70, 77, 84, 91, 98.$$

Checking these, we see that

$$21 = (2+1) \times 7, \ 42 = (4+2) \times 7, \ 63 = (6+3) \times 7 \text{ and } 84 = (8+4) \times 7$$

have the required property. There are therefore four such numbers.

Solution for Problem 1.7:

(a) Since the largest digit is 9, the number of digits in a number with a given digit sum will be the smallest if the number has as many nines as possible. Since $97 \div 9 = 10$ with remainder 7, such a number must have at least 11 digits. Since the remainder is 7, the smallest number with the given property is 79 999 999 999.

(b) This part is trickier. If the product of the digits of a number is equal to 29 400, each digit must be a divisor of 29 400. In order to write this number as the product of as few divisors as possible, we must consider its prime decomposition $29\,400 = 2^3 \times 3 \times 5^2 \times 7^2$. Both 5 and 7 are prime digits, none of whose multiples are single digits, and it therefore follows that there must be two fives and two sevens in any number with this digit product. The remaining divisor $2^3 \times 3 = 24$ is not a single digit, but can be written as the product of two single digits in two ways, namely $24 = 3 \times 8 = 4 \times 6$. The smallest number of digits any number with the required property can have is therefore six. In order for the lead digit to be as small as possible, we must choose the option $24 = 3 \times 8$, which yields 3 as a possible lead digit, and putting the digits in ascending order gives us 355778 as the smallest number with the required property.

Solution for Problem 1.8:

As was the case in the last problem, it will once again be useful to take a closer look at the prime decomposition of the given number. Since $2160 = 2^4 \times 3^3 \times 5$, we immediately see that 5 must be a digit of any number with the required property, as there is no single-digit multiple of 5. If the number we are seeking has three digits, the product of the other two must be $2^4 \times 3^3 = 432$, but this is not possible, since the largest possible product of two digits is $9 \times 9 = 81$. An integer of the given type can therefore not have only three digits.

We can, however, write $2^4 \times 3^3$ as the product of three digits, namely $6 \times 8 \times 9$. The smallest number of digits any number of the required type can have is therefore 4, and since the digits in this case are 5, 6, 8 and 9, putting them in ascending order gives us 5689 as the smallest number fulfilling the given condition.

Solution for Problem 1.9:
The two-digit number with the largest digits, and therefore the sum of whose digits is the largest, is 99. The largest sum possible is therefore $9 + 9 = 18$, and the largest rest that can result from such a division is therefore less than or equal to 17.

This rest is only possible if the divisor is 18, however, and the only number with digit sum 18 is 99. Dividing 99 by 18 does not leave this rest, however, since we have

$$99 = 5 \times 18 + 9.$$

We see that the rest 17 is not possible.

For numbers with digit sum 17, we could obtain the rest 16, but the only two numbers with digit sum 17 are 89 and 98, and since we have

$$89 = 5 \times 17 + 4 \quad \text{and} \quad 98 = 5 \times 17 + 13,$$

we see that the rest 16 is not possible either.

The next best option is the possible rest 15, which can turn up for division of a number with digit sum 16. Indeed, because

$$79 = 4 \times 16 + 15$$

holds, we see that 15 does indeed show up as a rest after division of the number 79 by the sum of its digits, and this is therefore the largest possible such a rest.

Solution for Problem 1.10:
The largest two-digit number is 99. Since this number also has the largest possible digits (i.e. both are nines), the sum of the digits is $9 + 9 = 18$, and the sum of the number with the sum of its own digits is equal to $99 + 18 = 117$. Since both the number and the digits are maximal, this is also the largest attainable sum.

The argument for the small case is similar. The smallest two-digit number is 10. The sum of its digits is $1 + 0 = 1$, and this is certainly the smallest possible value, since 0 is not attainable, as it would mean that all of the digits of a number are 0, which is only the case for 0 itself, which is not a two-digit number. Here, we therefore obtain the sum $10 + 1 = 11$. Since both the number and the digits are minimal, this is also the smallest attainable sum.

Solution for Problem 1.11:

First of all, we note that a number with the required property cannot have less than four digits. This is because the largest number with three digits is 999, and the largest possible sum that three digits can have is equal to $9 + 9 + 9 = 27$. It follows that the largest possible value obtained from a three-digit number (or even for a number with at most three digits) is equal to $999 + 27 = 1026$, which is quite a bit too small.

Note that the number 1026 is "almost" 2016; we only need to exchange the tens- and thousands-digits. This would seem to imply that the calculation leading to this specific result could be modified just a bit to obtain a solution. It isn't really immediately clear what this slight modification could be, though, so we will need to dive more deeply into this one.

So, what could be a next step? Well, we know that any number with the required property must have four digits. The first digit can obviously not be larger than 2, since the resulting sum would otherwise be too large. It follows that the first digit could only conceivably be 1 or 2.

What if the first digit is a 2? Since the number in question must be smaller than 2016, it can either be of the form 200* or 201*, with the star standing in for the ones-digit in each case. In the first case, the sum of the digits is $2 + 0 + 0 + {}^* = 2 + {}^*$, and adding the number to the sum of its digits gives us $200^* + 2 + {}^*$. Since this is to be equal to 2016, we have $^* + 2 + {}^* = 16$, or $^* = 7$, giving us the number 2007. This number does indeed have the required property, since the sum of its digits is $2 + 0 + 0 + 7 = 9$, and $2007 + 9 = 2016$. In the second case, the sum of the digits is $2 + 0 + 1 + {}^* = 3 + {}^*$, and adding the number to the sum of its digits gives us $201^* + 3 + {}^*$. Since this is to be equal to 2016, we have $10 + {}^* + 3 + {}^* = 16$. Unfortunately, this gives us $^* + {}^* = 3$, and this is not possible if * is a digit. There is therefore no number of this type with the required property.

Now, what if the first digit is a 1? Each of the other three digits must than be less than, or equal to, 9, and the sum of the digits is therefore at most $1 + 9 + 9 + 9 = 28$. The number we are looking for may therefore not be less than $2016 - 28 = 1988$. We therefore once again have two options to consider. The number can either be of the form 199* or 198*, with the star again standing in for the ones-digit in each case. In the first case, the sum of the digits is $1 + 9 + 9 + {}^* = 19 + {}^*$, and adding the number to the sum of its digits gives us $199^* + 19 + ^*$. Since this is to be equal to 2016 and because of the carry-over, we have $^* + 9 + ^* = 16$, which gives us $^* + ^* = 7$, which is again not possible if * is a digit, giving us no solution. In the second case, the sum of the digits is $1 + 9 + 8 + ^* = 18 + ^*$, and adding the number to the sum of its digits in this case gives us $198^* + 18 + ^*$. Since this is once again to be equal to 2016, we must have a carry-over from the ones-digit, and we therefore must have $^* + 8 + ^* = 16 + 10$, or $^* = 9$, giving us the number 1989. This number does indeed have the required property, since the sum of its digits is $1 + 9 + 8 + 9 = 27$, and $1989 + 27 = 2016$.

In summary, we see that there are two numbers with the required property, namely 1989 and 2007.

Solution for Problem 1.12:
In principle, the product of the digits of the number in question could be any power of three greater than or equal to 27. In fact, this will not apply, however. Since the product of the digits is not divisible by any prime other than three, this must also be true of each of the individual digits. We now note that there are only three digits not divisible by some other prime, namely 1, 3 and $9 = 3^2$. Any number with these three digits will indeed have $1 \times 3 \times 9 = 27$ as the product of its digits, but no other combination of digits is possible. It follows that the sum of the digits we are asked to determine is $1 + 3 + 9 = 13$.

Solution for Problem 1.13:
The sum of the digits of 24 is $2 + 4 = 6$. Any of its two-digit in-laws must therefore be multiples of 6. Since the divisors of 24 are 1, 2, 3, 4, 6, 8, 12 and 24 (yes, there are a lot of them), these numbers are candidates for the sums of digits of the in-laws of 24. We therefore only need to list the two-digit multiples of 6, and check to see in which cases the sums of their digits are numbers on this list.

The following table contains all two-digit multiples of 6 in the top row and the sums of their digits underneath them:

12	18	24	30	36	42	48	54	60	66	72	78	84	90	96
3	9	6	3	9	6	12	9	6	12	9	15	12	9	15

Checking the list of candidate sums (and noting that 24 cannot be its own in-law), we see that the in-laws of 24 are the seven numbers 12, 30, 42, 48, 60, 66 and 84.

Solution for Problem 1.14:

The key to solving this one is to realize that the common property of the elements of A gives us some extra information about divisibility. A number is divisible by 3 if and only if the sum of its digits is divisible by 3. Obviously, any number whose digits sum to 6 has this property, and it therefore follows that all numbers a in A are divisible by 3.

Having noticed this, the way forward would seem to indicate that it might be useful to consider the elements b of B for divisibility by 3. If we let t denote the tens-digit of some number b in B, we are given that the ones-digit is smaller by one, and must therefore be equal to $t - 1$. On the other hand, we are also given that the hundreds-digit is larger by one, and it must therefore be equal to $t + 1$. The sum of the digits of b is therefore equal to $(t + 1) + t + (t - 1) = 3t$, and since the sum of the digits of b is a multiple of three, we know that b itself must also be divisible by three.

In summary, this means that all elements of A are divisible by 3, as are all elements of B. No matter how we choose the numbers a and b, three will therefore always be a common divisor, and it is therefore not possible to choose numbers in the required way.

Solution for Problem 1.15:

Since an even number must have an even ones-digit, and the ones-digit of a palindrome is the same as the lead digit (i.e. the thousands-digit), which can never be 0, the only possible ones-digits are 2, 4, 6 and 8. On the other hand, any one of the ten digits from 0 to 9 can be the tens- (and therefore hundreds-) digit of such a number. Since these can be combined in any way, the total number of such numbers is equal to $4 \times 10 = 40$.

Solution for Problem 1.16:

If the first digit of the next such a number is again 3, the second digit must be at least 7. Since the five consecutive digits 3, 4, 5, 6, 7 cannot be used to create a number large than 37654 starting with the digits 3 and 7 (note that the digits 7, 6, 5 and 4 are in descending order!), the first digit of the next such a number cannot be 3, it must therefore be 4. The smallest five consecutive digits are 0, 1, 2, 3 and 4, and the smallest number starting with 4 and made up of these digits is 40123.

We see that Mr. Leadfoot will have to drive another $40123 - 37654 = 2469 \, \text{km}$ until he can enjoy such a display on his odometer again.

Solution for Problem 1.17:
As result of her multiplication, Caroline obtains a four-digit multiple of
9 with four consecutive increasing numbers as digits. Let t denote the
thousands-digit of the product. The hundreds-digit is then $t + 1$, the tens-
digit is $t + 2$, and the ones-digit is $t + 3$. Since the product is divisible by 9,
the sum of its digits, which is equal to $4t + 6$, which is obviously even and
less than 36, must be divisible by 9 as well. We therefore have $4t + 6 = 18$
and, finally, $t = 3$. This means that N certainly exists, with $9N = 3456$
and $N = 3456 \div 9 = 384$.

Solution for Problem 1.18:
If the ones-digit of a natural number n is divisible by 4, n is divisible by 4
if and only if its tens-digit is even. From this, we see that, if the ones-digit
of a natural number is equal to 8, its tens-digit must be even. (We already
know that the number is even, and the ones-digit can therefore not be 9.)
This means that the tens-digit of such a number is at most equal to 6, and
its hundreds-digit is at most equal to 5. Since 568 is divisible by 8, 568 is
a natural number divisible by 8 with ascending digits. Consequently, the
hundreds-digit of the largest of these numbers is not less than 5. On the
other hand, if the ones-digit of any even natural number with ascending
digits is less than 8, it is at most equal to 6, and the hundreds-digits of the
number is then certainly less than 5. The correct answer is therefore 5.

Solution for Problem 1.19:
If we consider one-, two- and three-digit numbers separately, we can first
note that there are 8 one-digit numbers with this property, namely all those
from 1 to 8. Advancing to the two-digit numbers, we note that each of the
two digits can similarly be any digit from 1 to 8, and that their choice is
completely independent. There are therefore $8 \times 8 = 64$ such numbers.
Also, considering the three-digit numbers with this property, we have the
same condition holding for three independent digits, which gives us a total
of $8 \times 8 \times 8 = 512$ such numbers. Since the only four-digit number 1000
in the span we are considering does not have the required property, the total
we are looking for is equal to $8 + 64 + 512 = 584$.

Solution for Problem 1.20:
First of all, it is clear that the Trenixians do not use the numbers divisible
by 3. From 1 to 100, there are $100 \div 3 = 33$ of these (not counting the rest

this division leaves, of course). The other kind of numbers they don't use, are those including the digit 3, and there are 19 of those between 1 and 100, namely those ending in 3, like 3, 13, 23, 33, 43, . . . , 93 and those beginning with 3, like 30, 31, . . . , 39. Since there are ten of each, that would seem to mean that there are 20 numbers written with the digit 3, but 33 has both of these properties, so we would have counted that one double.

It looks like we could just subtract the number of multiples of 3 and the number of numbers written with the digit 3 and be done with it, but there are numbers with both properties, of course, so we would be subtracting them twice. We can list these numbers explicitly: 3, 33, 63, 93, 30, 36 and 39. We see that there are 7 of these. This means that we obtain the number of numbers the Trenixians actually use in counting from 1 to 100 as

$$100 - 33 - 19 + 7 = 55,$$

and we see that the Grand Whoopty is actually turning 55.

Solution for Problem 1.21:
The first digit of any big four-digit number must be 9. We can use the digit 9 twice in A, since only three different digits are needed, so 9 can also be the hundreds-digit. The last two digits must be different, but as large as possible. The next largest digits are 8 and 7, and a number with the tens-digit 8 is bigger than one with tens digit 7 if all the digits to the left of it are the same. We see that $A = 9987$ must hold.

Similarly, the first digit of any small number must be 1. The next two digits can even be 0, which was not possible for the first digit. Finally, the last digit cannot be 0 or 1, so the smallest option is the digit 2. We therefore have $B = 1002$, and therefore $A - B = 8985$.

Solution for Problem 1.22:
The next smaller number before 999 that can be written with all the digits the same is 888. This means that the last year before $2 \times 999 = 1998$ with the property that it has a number twice that of one written with all the digits the same is $2 \times 888 = 1776$.

The other part of this problem is a bit trickier. Since there is no digit larger than 9, we have to come up with some other way to raise the value of the number, and that means increasing the number of digits from three to four. The smallest digit is 0, but the leading digit of a number can't be 0, so 0000 is not a candidate. (Also, it wouldn't be bigger than 999 anyway.) So, the next possible number is written with all ones, making 1111 the next largest number after 999 to be written with all digits the same. This means that the next year after $2 \times 999 = 1998$ with the required property is $2 \times 1111 = 2222$.

The number of year that have passed between those two is, of course,

$$2222 - 1776 = 446.$$

Solution for Problem 1.23:
Any three-digit number written with exactly two different digits a and b must be of one of the three forms aab, aba or abb. In each of these cases, a can be any one of the nine digits from 1 to 9 and b can be any one of the nine digits from 0 to 9 not equal to a. This means that there are $9 \times 9 = 81$ numbers of each type, and therefore $3 \times 81 = 243$ numbers of the required type altogether.

Solution for Problem 1.24:
The smallest possible product of two two-digit numbers is $10 \times 10 = 100$, and this number has three digits. The largest possible such product is certainly smaller than $100 \times 100 = 10000$, and must therefore have less than five digits. It follows that a number of the required type can only be composed of either three or four digits.

Let us assume that 888 is the product of two two-digit numbers. In order to find these factors, it is useful to consider its prime decomposition $888 = 2^3 \times 3 \times 37$. Since 37 is a prime, one of the factors must be a two-digit multiple of 37. The only options for such a factor are 37 itself and $2 \times 37 = 74$, since $3 \times 37 = 111$ is already too large. This gives us two possible combinations of two-digit numbers whose product equals 888, namely $37 \times (2^3 \times 3) = 37 \times 24$ and $74 \times (2^2 \times 3) = 74 \times 12$.

Now, let us assume that 8888 is the product of two two-digit numbers. Once again, we consider its prime decomposition $8888 = 2^3 \times 11 \times 101$. Since 101 is a prime, we see that one of these two-digit numbers must be a multiple of 101, which is not possible, since 101 already has three digits on its own. No such a combination of two-digit numbers can therefore exist.

In summary, we see that there are two combinations of two-digit numbers that yield a product of the desired type, namely 37×24 and 74×12.

Solution for Problem 1.25:

All two-digit numbers have this property, with the exception of multiples of 11, i.e. 11, 22, 33, ..., 99. Since there are $99 - 9 = 90$ two-digit numbers in all, of which 9 are multiples of 11, we have $90 - 9 = 81$ two-digit numbers with the desired property.

Turning our attention now to three-digit numbers of the desired type, we first notice that their hundreds digit must be 1, since they are to be smaller than 200. There are three types of three-digit numbers starting with 1 and containing exactly two different digits 1 and a, namely numbers of the types $11a$, $1a1$ and $1aa$. Since a can be any of the ten digits from 0 to 9 with the exception of 1, this gives a total of $3 \times 9 = 27$ numbers of this type.

Altogether, there are therefore $81 + 27 = 108$ numbers with the desired properties.

Solution for Problem 1.26:

Any number composed of k times the digit 9, is one less than the kth power of ten, since this number is simply written as a 1 followed by k zeros. This means that we can write

$$a = 10^{20} - 1 \quad \text{and} \quad b = 10^{12} - 1.$$

Their product can therefore be written as

$$a \times b = (10^{20} - 1) \times (10^{12} - 1),$$

and some simple calculation gives us

$$(10^{20} - 1) \times (10^{12} - 1) = 10^{32} - 10^{20} - 10^{12} + 1.$$

This may seem like a really complicated expression, but the structure of this number is really quite simple when you take a closer look. Just consider part of the number, i.e. $10^{32} - 10^{12}$. What happens when you subtract a smaller power of ten from a larger one? Consider the expression $10^5 - 10^2$. This can also be written in the following way:

$$10^5 - 10^2 = 100000 - 100 = 99900.$$

If we describe this in words, we can say that we subtract the number with a 1 followed by 2 zeros from the number with a 1 followed by 5 zeros, and obtain a number made up of 3 nines and 2 zeros. The three results from $5 - 2 = 3$. By way of analogy, it should now be clear that the result of

$10^{32} - 10^{12}$, i.e. subtracting the number with a 1 followed by 12 zeros from the number with a 1 followed by 32 zeros, results in a number made up of $32 - 12 = 20$ nines and 12 zeros. Subtracting 10^{20} (i.e. a 1 followed by 20 zeros) from this number will just reduce the 21st digit from the end by 1, and thus change one of the nines to an eight. Similarly, adding 1 to the number will simply raise the final digit from 0 to 1.

This means that the result of the calculation is a number composed of 19 nines, one eight, 11 zeros and a one. It is therefore not difficult to calculate the sum of these digits as

$$19 \times 9 + 1 \times 8 + 11 \times 0 + 1 \times 1 = 171 + 8 + 0 + 1 = 180.$$

Solution for Problem 1.27:
If a number is divisible by 9, the sum of its digits must also be divisible by 9. Since the number *babab* is divisible by 9, so is $3b + 2a$. We also know that *babab* is not divisible by 18, but since it is divisible by 9, this means that it is not divisible by 2. In other words, we know that *babab* is odd, and therefore its ones-digit *b* must also be odd. Furthermore, since $3b + 2a$ is divisible by 9, it is certainly divisible by 3, and this means that *a* must also be divisible by 3. We also know that N is divisible by 6. This means that N must be even, and therefore its ones-digit *a* must also be even. We see that *a* can only be equal to 6, since it cannot be 0 as the lead digit of N.

The only odd value of *b*, such that $3b + 2a = 3b + 12$ is divisible by 9, is therefore 5. (Note that we have obtained $N = 6565656$.)

Solution for Problem 1.28:
Let us call the number we obtain after removing the first digit six from the original given number. Tacking the 6 onto the end of this number therefore results in the number $10x + 6$. We want this x to be as small as possible. This means that it should have as few digits as possible, and we can just check to see what happens if x has 0, 1, 2, ... digits.

Obviously, x cannot have 0 digits, as the two numbers we are considering would then be the same, namely 6. If x has one digit, the original number is $60 + x$, and the given condition gives us

$$60 + x = 4 \times (10x + 6) \quad \text{or} \quad 60 + x = 40x + 24 \quad \text{or} \quad 36 = 39x,$$

which is obviously not possible.

If x has two digits, the original number is $600 + x$, and we have

$$600 + x = 4 \times (10x + 6) \quad \text{or} \quad 600 + x = 40x + 24 \quad \text{or} \quad 576 = 39x,$$

and this does not give us an integer value for x.

If x has three digits, the original number is $6000 + x$, and we have

$$6000 + x = 4 \times (10x + 6) \quad \text{or} \quad 6000 + x = 40x + 24 \quad \text{or} \quad 5976 = 39x,$$

and this does not give us an integer value for x either.

If x has four digits, the original number is $60000 + x$, and we have

$$60000 + x = 4 \times (10x + 6) \quad \text{or} \quad 60000 + x = 40x + 24 \quad \text{or}$$
$$59976 = 39x,$$

and this again does not give us an integer value for x.

However, if x has five digits, the original number is $600000 + x$, and we have

$$600000 + x = 4 \times (10x + 6) \quad \text{or} \quad 600000 + x = 40x + 24 \quad \text{or}$$
$$599976 = 39x,$$

and this gives us the smallest possible integer value $x = 15384$.

We see that the smallest possible number with the property described is 615384.

Solution for Problem 1.29:
Since the resulting number is to be divisible by 5, its last digit must be 5, as there is no digit 0 in the given number. This means that we certainly have to delete the final digit 1 of the number. This gives us the number 467658425.

We now have four more digits to delete. We want the result to be as large as possible, and this means that the lead digit should be as large as possible. It cannot be 9, since there is no digit 9 in the given number. It cannot be 8 either, since the only digit 8 in the given number is the sixth digit, and it can only advance to the front if we delete all five digits to its left. This is not possible, however, since we only have four digits left to delete. The largest possible lead digit of the resulting number after deletion is therefore 7. In order for 7 to be the lead digit, we must delete the two digits to its left. This gives us the number 7658425, leaving us two more digits to delete.

Now, we can use the 8 as the second digit of the resulting number. It is the largest of the remaining digits, and there are only two digits between the leading digit 7 and this digit. We therefore delete the digits between the 7 and the 8, and obtain 78425 as the largest possible result.

Since we deleted the digit 1 in the first step, the digits 4 and 6 in the second, and the digits 6 and 5 in the third, the sum of the deleted digits is equal to

$$1 + 4 + 6 + 6 + 5 = 22.$$

Solution for Problem 1.30:

If a number is divisible by 5, its ones-digit must be either 0 or 5. We know that the number we are searching for has the same first and last digit, and since the first digit of a number can never be 0, this common first and last digit must therefore be 5. Since we are also given that the hundreds digits is a 6, the number must be of the form $5a6b5$, for some digits a and b.

We are also given that the number $a6b$ is divisible by 4. A number is divisible by 4 if and only if the number formed by its last two digits is divisible by 4. In this case, it means that the number $6b$ is divisible by 4, and it therefore follows that b must be either 0, 4 or 8, since 60, 64 and 68 are the two digits numbers starting with the digit 6 that are divisible by 4.

We therefore have three options for the number we are searching for, namely $5a605$, $5a645$ and $5a685$. We can now apply the last bit of information we were given, namely that the number is also divisible by 9. A number is divisible by 9 if and only if the sum of its digits is equal to 9. For a number of the form $5a605$, this means that $5 + a + 6 + 0 + 5 = 16 + a$ is divisible by 9, and this is only possible for $a = 2$, which yields the first number with the required properties as 52605. For a number of the form $5a645$, this means that $5 + a + 6 + 4 + 5 = 20 + a$ is divisible by 9, and this is only possible for $a = 7$, which yields the second number with the required properties as 57645. Finally, for a number of the form $5a685$, this means that $5 + a + 6 + 8 + 5 = 24 + a$ is divisible by 9, and this is only possible for $a = 3$, which yields the third and final number with the required properties as 53685.

Solution for Problem 1.31:

If we want the product of the two-digit numbers to be as large as possible, we must choose the tens-digits of these numbers as large as possible. This means that the 8 and the 7 must be the tens-digits of the two numbers, and this leaves us only two options for the numbers. Checking the possible products, we obtain

$$85 \times 72 = 6120 \quad \text{and} \quad 82 \times 75 = 6150,$$

and the largest possible product we are searching for is therefore 6150.

Solution for Problem 1.32:
Charlie's notation always gives the digit followed by the number of times in a row the digit is written, separated by a _. His way of writing 4_15_36_2 + 4_56_1 can therefore be interpreted in the following way:

$$4\_15\_36\_2 + 4\_56\_1 = 455566 + 444446 = 900012 = 9\_10\_31\_12\_1.$$

Solution for Problem 1.33:
If N is divisible by 4, it must certainly be even, and this means that its ones-digit must be even. We now note that a natural number is divisible by 4 if and only if the two-digit number formed by its tens-digit and its ones-digit is divisible by 4.

Suppose the ones-digit of N is equal to 2. Then it is a necessary and sufficient condition for the tens-digit of N to be odd. Consequently, 1, 3 and 5 are all possible tens-digits of N.

Suppose the ones-digit of N is equal to 4. Then, the tens-digit of N must be even, leaving 2 as the only possible tens-digit in this case, since each of the five digits from 1 to 5 can only be used once.

In summary, this means that the tens-digit of N cannot be equal to 4.

Solution for Problem 1.34:
This is easier if you consider a smaller version. If you subtract 1 from 10^4, you obtain

$$10^4 - 1 = 10000 - 1 = 9999.$$

Put into words, subtracting 1 from a number with a one followed by four zeros results in a number composed of four nines.

The analogous results is true for any power of ten. Subtracting 1 from a number with a one followed by n zeros results in a number composed of n nines, and specifically, subtracting 1 from a number with a one followed by 100 zeros results in a number composed of 100 nines.

We can now note that subtracting 1994001994 from some number can also be done by first subtracting 1 and then subtracting 1994001993 in a second step. We already know that the result of the first subtraction gives us a number composed of 100 nines. Subtracting 1994001993 from this number does not affect the first 90 digits at all, but the last ten digits of the final result are then given by

$$9999999999 - 1994001993 = 8005998006.$$

There are only two nines left in this part, and together with the 90 nines to the left of it, we see that the result of the subtraction is a number with $2 + 90 = 92$ nines.

Solution for Problem 1.35:
First of all, let us see how we can obtain a maximum difference. We want to subtract something very small from something very large. Obviously, the largest possible numbers will start with large digits, and the largest possible are two nines. The smallest possible will start with small digits, and the smallest possible are zero and one. The best candidate for a maximum difference therefore appears to be

$$9901 - 1066 = 8835.$$

This is indeed the best we can do, since any combination of lead digits other than 99 and 10 will certainly yields lead digits less than 88 in the difference.

Now, what about minimizing the difference. This is a whole different kettle of fish. Of course, the smallest possible difference is zero, since we could consider $8888 - 8888$ or $1111 - 1111$, each of which would yield 0, as well as several other numbers that don't change when you flip them, like 1961 or 1881, and you get full points for finding this answer.

The more interesting question to ponder here however is to find the smallest possible positive difference we can get. How can we minimize the difference under the assumption that the number changes when we flip it?

Considering the cases of 1111 and 8888 does give us the idea that we could use 1 or 8 both as the first and last digits, because this will certainly result in a difference with less than four digits. A good option seems to be

$$1981 - 1861 = 120,$$

but how do we decide whether this is the smallest we can get?

Let us consider a number *abcd*. If the first and last digit do not flip to each other, the difference will have a thousands-digit that can only be eliminated by a carry-over. The best we can then do in the hundreds-digit is then $0 - 9$, but this results in $6 - 0$ in the tens-digit by flipping, and the difference will therefore certainly have at least three digits. We can, however attain a difference with two digits by letting the first and last digits flip to each other, and keeping the hundreds- and tens-digits separated by only one. There are only two combinations that could make this possible, namely $1 - 0$ and $9 - 8$. Our best options are therefore the already suggested $1981 - 1861 = 120$ (which is also attainable with other first-last digit combinations, like $6989 - 6869$, for instance, and does not yield a two-digit difference) and $1101 - 1011 = 90$ (which is again attainable in other ways, like $9106 - 9016$). This is actually the only way to obtain a two-digit difference, and the smallest possible difference is therefore 90.

Chapter 2

A Little Bit of Algebra — Word Problems and Variables

There are problems for which it can be really useful to introduce a variable and solve an equation. That doesn't necessarily mean that this is the only way to proceed, but a little bit of simple algebra can often make life easier. Often the advantage just lies in the fact that a complicated text is really not so complicated if you can write it using variables as short-cuts.

See if you can find the short-cuts that can help find the solution to these problems. If you don't see them right away, don't forget that you can check out the tips at the end of the list to help you out.

PROBLEMS

Problem 2.1:
I was once confronted with a fascinating five-digit number. I called it A. If I added the digit 1 at the start of the number, I obtained a six-digit number. If I added the digit 1 at the end, I also obtained a six-digit number, but this one was exactly three times as large as the first one. What was this number A?

(StU-89-4)

Problem 2.2:
A folk dancing group is practising a slap dance, in which all the members are arranged in a big circle. During the course of the dance, each dancer gets slapped 5 times by each of his or her direct neighbors and twice by their neighbors' neighbors in the circle. (They do not slap themselves, however.)

31

They also get slapped once by every other dancer. After the dance is over, each dancer has been slapped 16 times. How many dancers are there?

(O-96-1)

Problem 2.3:
A book store sells a lot of copies of a new book. On the first day, it sells one eighth of its stock plus 10 books. On the second day, it sells half of the remaining stock plus 15 books. After this, there are still 50 books in stock. How many did the book store have in stock at the beginning?

(StU-93-13)

Problem 2.4:
A certain number of questions of equal value are posed on a test. If I answer 9 of the first 10 correctly, and exactly $\frac{3}{10}$ of the remaining ones, I will have exactly 50% of the answers correct. How many questions are posed on the test?

(O-93-5)

Problem 2.5:
Franz has just returned from a shopping expedition. He spent one eighth of his money in the first store he visited. He then spent 50 euros less in the second store than he did in the first. In the third store, he spent twice as much as he did in the second store, and he then spent 300 euros in the fourth store. This left him with 150 euros more in his wallet than he has spent at the first store. How much money did he start out with at the beginning of his expedition?

(StU-18-B3)

Problem 2.6:
Karl, his mother and his sister Anna are 56 years old in total. When Karl was as old as his sister is now, his mother was 32 years old. How old is Karl now?

(O-03-4)

Problem 2.7:
Mr. Dreher's age together with that of his wife will be equal to 108 when his wife is as old as he is now. How old is Mr. Dreher and how old is his

wife if we know that Mr. Dreher is now twice as old as his wife was when he was the age his wife is now?

(StU-92-9)

Problem 2.8:
Andreas is observing a flock of birds. One seventh of them lands in the branches of an oak, one third of them lands in a meadow, twice the difference between these two numbers land on the branches of a fir tree and six are still flying around, since they cannot decide whether they should head for the white blossoms of the apple tree or the pink blossoms of the aromatic peach tree. How many birds are there in the flock?

(StU-97-B4)

Problem 2.9:
A farmer trades cheese for chickens, and two wheels of cheese are equivalent to three chickens. The chickens lay eggs, and every day each of them lays one third as many eggs as there are chickens. The farmer brings these eggs to the market, and sells each carton of nine eggs for three times as many Schillings as a chicken lays eggs in one day. On this day, the farmer has earned 216 Schillings. How much cheese did the farmer originally trade for chickens?

(StU-92-6)

Problem 2.10:
Farmer Berger and his hired hand Sepp are trying to weigh a calf, but it refuses to hold still. In order to get its weight, Sepp holds on to the calf and gets on the scale with it. Farmer Berger reads their weight as 98 kg. Next, Farmer Berger holds the calf while Sepp reads his weight with the calf as 107 kg. Finally, Sepp holds Farmer Berger, and the calf reads their weight as 153 kg. How much does the incredibly intelligent (but fidgety) calf weigh?

(StU-97-A3)

Problem 2.11:
Selina is given two positive integers a and b and calculates the values of $a + b$, $a - b$, $a \times b$ and $a \div b$. She writes the four results on four pieces of paper. After a while, she has forgotten the original numbers and misplaced the papers. On finding two of them, she discovers the numbers 2 and 15

written on them. Prove that these cannot be the values of $a + b$ and $a - b$ and determine all possible combinations of a and b that might have led to these values.

(StU-15-B1)

TIPS

Tip for Problem 2.1:
If you add a 1 at the beginning of a two-digit number (like 47, for instance), you turn the number 47 into 147. In other words, you add 100 to the two-digit number. If you add it at the end, you turn 47 into $471 = 470 + 1 = 47 \times 10 + 1$. In other words, you multiply the two-digit number by 10 and add 1. What happens if you do the same with a five-digit number?

Tip for Problem 2.2:
If there are d dancers in the circle, you can express the number of times each dancer gets slapped in terms of d.

Tip for Problem 2.3:
You could try starting with the idea that there were just x books in stock at the beginning and go on from there, if you like algebra. You could also try figuring this one out backwards, though!

Tip for Problem 2.4:
If you just have the idea of replacing the number of problems on the test with a variable, it shouldn't be hard to write this as an equation.

Tip for Problem 2.5:
This is one of those problems where the obvious way to go will work. What is it you want to find out? Call it x!

Tip for Problem 2.6:
It seems like a pretty obvious idea to use a variable to represent one of the ages in this problem. Still, it isn't so easy to represent all the other information in terms of that variable right away. How about another variable? How much older than Anna is Karl, for instance?

Tip for Problem 2.7:
If you solved Problem 2.6, this one should be easy. You can use exactly the same idea.

Tip for Problem 2.8:
Don't let the fractions bother you! You know what you want to find out. Call it x and you'll be fine.

Tip for Problem 2.9:
This kind of problem seems really complicated when you first read it, but translating the text into an algebraic expression can make it a lot easier. There are a lot of things we don't know the value of, and we can use a variable to represent any one of them. As a next step, we can then translate the text bit by bit into expressions we can calculate with directly.

Tip for Problem 2.10:
First of all, you're going to have to get over the fact that the calf can read a scale. He really is that smart. Once you are able to accept this, you might try to figure out what the total weight of three of them (Farmer Berger, Sepp and the calf) is. Once you get that far, you're practically done.

Tip for Problem 2.11:
Look at the sum of these two numbers. Notice anything special?

SOLUTIONS

Solution for Problem 2.1:
Adding a 1 at the start of a five-digit number is the same as adding 100 000 to the number, and adding a 1 at the end is the same as multiplying the number by 10 and then adding 1.

Writing this in an algebraic way, the first operation turns the number A into $A + 100\,000$, and the second turns A into $10A + 1$. We are told that the second number is three times as large as the first. This means that we can write

$$10A + 1 = 3 \times (A + 100\,000).$$

With simple algebraic manipulation, this is the same as writing

$$10A + 1 = 3A + 300\,000, \quad \text{or} \quad 7A = 299\,999.$$

Dividing by 7, we see that the fascinating number A was 42 857.

Solution for Problem 2.2:

The key to this problem is figuring out how many dancers there are in the circle that slap a given dancer exactly once.

Imagine that you are one of the d dancers in the circle. You do not slap yourself. You get slapped 5 times by each of your two neighbors and twice by each of the two neighbors' neighbors, and this leaves $d - 5$ other dancers in the circle to slap you one time each.

You therefore get slapped a total of $2 \times 5 + 2 \times 2 + (d - 5)$ times, and since we are given that this number is equal to 16, we obtain the equation

$$2 \times 5 + 2 \times 2 + (d - 5) = 16.$$

Since this reduces to $10 + 4 + d - 5 = 16$, we obtain $d = 7$ dancers in the circle.

We can check this by noting that there are now two dancers who you will get slapped by once. You are therefore slapped a total of $10 + 4 + 2 = 16$ times, as was stated.

Solution for Problem 2.3:

If you had a good look at the tips before you tried this one, you saw that there are two good ways to look at this. If we let x stand for the number of books in stock at the beginning, the store sells $\frac{1}{8} \times x + 10$ books the first day, leaving $x - (\frac{1}{8} \times x + 10) = \frac{7}{8} \times x - 10$ books in the store. Since half of these are sold the next day, and then another 15, leaving 50 in stock, we obtain the equation

$$\frac{1}{2} \times \left(\frac{7}{8} \times x - 10 \right) - 15 = 50.$$

This is easy to solve, and we get $x = 160$.

On the other hand, we can also get this answer by working backwards.

Since there were 50 books left after the second day, and we know that half of the stock plus 15 books were sold that day, there were $50 + 15 = 65$ books in stock before the last ones were sold, and since this was half of the stock after the first day, there were twice as many, or 130, left after the first day. This was the result of selling one eighth of the original stock plus ten books. Selling one eighth of the original stock therefore left 140 books, and since this was seven-eighths of the original stock, there were originally eight-sevenths as many, or 160, in stock.

Solution for Problem 2.4:

If we assume that there were t questions on the test, the number of correct answers was

$$9 + \frac{3}{10} \times (t - 10).$$

Since this number equals exactly half of the number of questions, we obtain the equation

$$9 + \frac{3}{10} \times (t - 10) = \frac{1}{2} \times t,$$

which we can write as $9 + \frac{3}{10} \times t - 3 = \frac{5}{10} \times t$, or $6 = \frac{2}{10} \times t$, which gives us $t = 30$.

We can check to see that this is correct. After the first 9 correct answers, $\frac{3}{10}$ of the remaining 20 questions were answered correctly. This this gives us a total of $9 + 6 = 15$ correct answers out of 30 questions, we see that exactly 50% were answered correctly.

Solution for Problem 2.5:

If Franz started out with x euros, he spent $\frac{x}{8}$ at the first store. Since he spent 50 euros less than that at the second store, this means that he spent $\frac{x}{8} - 50$ there. He doubled that at the third store, which means that he spent $2 \times (\frac{x}{8} - 50) = \frac{x}{4} - 100$ there. Since we know that he spent 300 euros at the fourth store, we can sum up his purchases, and this gives us the equation

$$\frac{x}{8} + \left(\frac{x}{8} - 50\right) + \left(\frac{x}{4} - 100\right) + 300 = x - \left(\frac{x}{8} + 150\right),$$

since we know that he was left with 150 euros more than the amount he spent at the first store. All that is left for us to do is to solve this equation, and since we can write it as

$$\frac{x}{2} + 150 = \frac{7x}{8} - 150$$

or $\frac{3x}{8} = 300$, we obtain $x = 800$ as our answer.

As usual, we can check whether this amount actually works. Since he spent one eighth at the first store, he spent 100 euros there. He spent 50 euros less, i.e. 50, at the second store, and twice that, i.e. 100, at the third. Altogether, he therefore spent $100 + 50 + 100 + 300 = 550$, which left

$800 - 550 = 250$ euros in his wallet, which is indeed 150 euros more than the 100 euros he spent at the first store.

Solution for Problem 2.6:
We can say that Anna is a years old. Since Karl is older, he is $a + x$ years old. This means that Karl was a years old x years ago, and his mother was 32 then. That means that she is $32 + x$ years old now.

We know that the sum of the ages of Karl, Anna and their mother is 56. This means that the equation

$$(a + x) + a + (32 + x) = 56$$

must hold. This can also be written as $2a + 2x = 24$ or $a + x = 12$. We already know that Karl is $a + x$ years old, and he is therefore 12.

Solution for Problem 2.7:
If we let x stand for Mr. Dreher's age and y for the number of years Mr. Dreher is older than his wife, his wife is now $x - y$ years old.

When his wife is as old as Mr. Dreher is now, his age will be equal to $x + y$, and their total age will be equal to $(x + y) + x = 2x + y$. We are told that this total will be equal to 108, and this gives us the equation

$$2x + y = 108,$$

which we can also write as $y = 108 - 2x$.

Mr. Dreher was the age his wife is now, namely $x - y$, y years ago. His wife was then

$$(x - y) - y = x - 2y$$

years old, and we are told that Mr. Dreher is now twice this age. This gives us another equation, namely

$$x = 2 \times (x - 2y).$$

Substituting $y = 108 - 2x$ in this equation gives us

$$x = 2 \times (x - 2 \times (108 - 2x)),$$

which simplifies to $x = 2x - 432 + 8x$ or $9x = 432$. We therefore see that $x = 48$, and Mr. Dreher is therefore 48 years old. Since $y = 108 - 2x$, we have $y = 108 - 96 = 12$, and Mr. Dreher is 12 years older than his wife.

She is therefore 36 years old. (Note that his wife was 24 when he was 36, and he is now twice that age, as was stated.)

Solution for Problem 2.8:

We want to know the number of birds in the flock, and there is no reason why we shouldn't just name that unknown number x. One seventh of x is $\frac{x}{7}$ and one third of x is $\frac{x}{3}$, while twice the difference of these two numbers is $2 \times (\frac{x}{3} - \frac{x}{7})$. Collecting the information given in the text, we can therefore add up the groups and obtain the following expression for the total number x of birds in the flock:

$$x = \frac{x}{7} + \frac{x}{3} + 2 \times \left(\frac{x}{3} - \frac{x}{7}\right) + 6.$$

Some easy simplification gives us

$$x = x - \frac{x}{7} + 6,$$

or $\frac{x}{7} = 6$, which gives us $x = 42$.

Sure enough, if there are 42 birds in the flock, one seventh of them (i.e. 6) land in the oak, one third (i.e. 14) land in the meadow, which twice the difference (i.e. 16) land in the fir tree, leaving 6 of the 42 to fly around.

Solution for Problem 2.9:

There are lots of things we don't know the value of, like the number of wheels of cheese, the number of chickens or the price of a carton of eggs. We could replace any of these with a variable, but let's see what happens if we let c represent the number of chickens.

Since two wheels of cheese are equivalent to three chickens, this means that the farmer originally traded $\frac{2c}{3}$ wheels of cheese for chickens.

We are told that each chicken lays one third as many eggs as there are chickens every day, which we can write as $\frac{c}{3}$, and since each of the c chickens lays this many, there are

$$\frac{c}{3} \times c = \frac{c^2}{3}$$

eggs laid every day. Since each carton contains nine eggs, the farmer has

$$\frac{1}{9} \times \frac{c^2}{3} = \frac{c^2}{27}$$

cartons to sell. Since each of these sells for a price equal to three times as many Schillings as a chicken lays eggs in one day, the farmer earns a total of

$$\frac{c^2}{27} \times \left(3 \times \frac{c}{3}\right) = \frac{c^3}{27}$$

Schillings, which we are told is equal to 216. This gives us the equation

$$\frac{c^3}{27} = 216.$$

Now, it is very helpful to know a little bit about number powers. We see that there is a third power of c in the mix here. But since $27 = 3^3$, the number 27 is also a third power. This is also true for 216, since $216 = 6^3$. We can therefore write

$$\frac{c^3}{3^3} = 6^3, \quad \text{or} \quad \frac{c}{3} = 6.$$

The number c of chickens is therefore equal to 18, and since we know that the farmer originally brought two-thirds this number of wheels of cheese to the market, the number we are looking for is equal to 12.

Solution for Problem 2.10:
Let us write the weight of Farmer Berger as b, the weight of Sepp as s and the weight of the calf as c.

We know from Farmer Berger's reading that $s + c = 98\,\text{kg}$. From Sepp's reading, we have $b + c = 107\,\text{kg}$, and from the calf's surprisingly useful reading, we have $s + b = 153\,\text{kg}$. Adding all three of these equations, we get

$$(s + c) + (b + c) + (s + b) = 98 + 107 + 153,$$

which we can also write as $2(s + b + c) = 358$, or $s + b + c = 179$. This makes it easy to figure out the calf's weight, since we have

$$c = (s + b + c) - (s + b) = 179 - 153 = 26\,\text{kg}.$$

Solution for Problem 2.11:
This is quite a different type of algebraic problem from the ones we have been looking at previously. This time, the variables simply stand for integers, but do not have any meaning beyond that. Also, we aren't able to reduce the problem to any simple equation. The argument will require some slightly more sophisticated reasoning.

If 2 and 15 are $a + b$ and $a - b$ in some order, then we have $2 + 15 = (a + b) + (a - b) = 2a$. Unfortunately, $2 + 15 = 17$ is an odd number, whereas $2a$ is certainly even, no matter which integer a happens to be. We see that the two numbers 2 and 15 cannot be the sum and the difference of the two numbers.

In order to find all possible combinations, let us now assume that $a \geq b > 0$ holds.

If $a + b = 2$, we have $a = b = 1$, and the number 15 cannot be the product, the quotient or the difference of a and b. This case is therefore not possible.

If $a \times b = 2$, we have $a = 2$ and $b = 1$, and once again the number 15 cannot be the product, the quotient or the difference of a and b. This case is therefore not possible either.

Now, what if $a - b = 2$ holds? We already know that 15 cannot be the sum of a and b. Can we have $15 = a \div b$? This is not possible either, since this would imply

$$15 = a \div b = (b + 2) \div b = 1 + \frac{2}{b} < 2.$$

The only option left for 15 is therefore $a \times b = 15$. This is equivalent to $(b + 2) \times b = 15$, and we therefore obtain $b = 3$ and $a = 5$ as the only possible combination in this case.

Finally, we still have the case $a \div b = 2$ left to consider. This can also be written as $a = 2b$. Assuming $a + b = 15$ therefore yields $2b + b = 3b = 15$, and therefore $b = 5$ and $a = 10$.

Assuming $a - b = 15$ yields $2b - b = b = 15$ and $a = 30$.

And that just about sums things up, since the odd product $a \times b = 15$ isn't possible for any even value of $a = 2b$.

We see that there are three pairs of numbers a and b that might have resulted in the numbers 2 and 15, namely (5, 3), (10, 5) and (30, 15).

Chapter 3

Building with Bricks and Blocks

Thinking in three spatial dimensions does not come easily to everyone. A fun way to approach this type of question is through problems concerning things built out of rectangular bricks. Sometimes the bricks are even cubic blocks, with all sides of the same length. Here are a few problems dealing with objects that are either built up out of rectangular bricks or cut up into such bricks.

PROBLEMS

Problem 3.1:

The two objects shown on the right are made up of four and three cubes, respectively, all of the same size. Which of the following objects cannot be formed by placing the two given objects together in an appropriate manner?

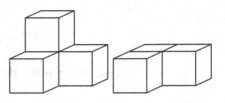

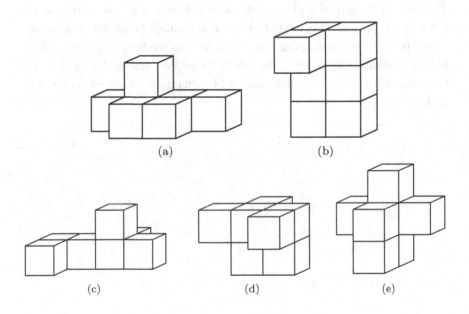

(a)

(b)

(c)

(d)

(e)

(StU-03-A3)

Problem 3.2:

We are given a $3 \times 3 \times 3$ cube, made up of smaller cubes. Holes with 1×1 square faces are cut in the middle of this cube in each direction, leaving the "skeleton" shown in the figure. This skeleton is then painted all over. How many 1×1 squares have to be painted?

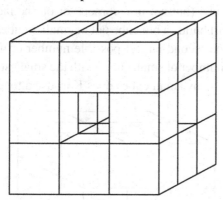

(StU-97-A4)

Problem 3.3:

The object in the figure is made up of four cubes of the same size. Its volume is $32\,\text{cm}^3$. Determine its surface area.

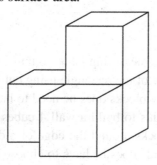

(StU-00-A8)

Problem 3.4:

If a big cube is built out of a number of smaller cubes in the manner shown in the figure, a small cube X in the interior of the big cube has a common face with six other small cubes. We can say that X has 6 neighbors. How many neighbors does a small cube have, if it lies in a corner of the cube, or at an edge of the cube (but not in a corner), or on the face of a cube (but not at an edge)? Karl builds a big cube out of smaller cubes. The number of small cubes with the second-largest possible number of neighbors is three times as big as the number of small cubes with the smallest possible number of neighbors. How many small cubes does Karl use altogether?

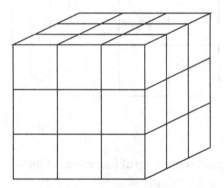

(StU-05-B2)

Problem 3.5:

Christian has a rectangle measuring 14 × 16 units. He builds a wall 1 unit thick all around the edge of the rectangle using cubic blocks with edges of length 1 unit. How many blocks does he need to build a wall 5 cubes high all around? He now wants to build a wall 4 cubes high all around, using the same number of blocks around the edge of a different rectangle with length 21 units. How wide does he have to choose this rectangle? Finally, he wants to rebuild the original wall surrounding the first rectangle with bricks measuring $3 \times 1 \times 1$ units instead of cubic blocks, but he finds that he cannot do this. Why not?

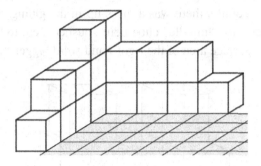

(StU-99-B1)

Problem 3.6:
A wooden cube with edges of length 5 cm is painted green. The cube is then cut up into small cubes with edges of length 1 cm. How many of the small cubes have exactly two green sides?

(O-97-2)

Problem 3.7:
Romana is building a solid rectangular block out of cubes with edges of length 1 unit. The length of the block is going to be 8 units and its width will be 10 units. For the outside of the block (i.e. the upper and lower layers, the front and back, and the leftmost and rightmost layers), she intends to use red cubes exclusively. For the interior, she intends to use blue cubes exclusively. In total, she will end up using the same number of red and blue cubes. What will be the height of the completed rectangular block?

(StU-12-B2)

Problem 3.8:

In the figure, you can see two solids, each of which has been built by gluing three cubes with edge lengths of 6 cm together. While the solid on the left was built very carefully, there was a mess-up in the gluing process on the right, which led to the "middle" cube being pushed 1 cm to the side. How many cm² is the surface area of the right-hand solid bigger than the one on the left?

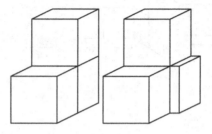

(S5K-05-6)

Problem 3.9:

We are given a big $5 \times 5 \times 5$ cube made up of 125 small $1 \times 1 \times 1$ cubes. Holes are made in the big cube by removing straight rows of small cubes from left to right, top to bottom, and front to back as shown in the figure, leaving a "skeleton" of the original big cube. How many of the original $1 \times 1 \times 1$ cubes are still left in this skeleton?

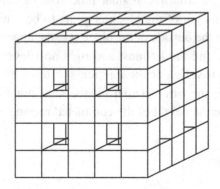

(S5K-06-8)

Problem 3.10:
A solid cube with sides of length 1 m is cut up into small cubes with sides of length 5 cm each. These small cubes are laid side by side to a long bar as shown in the figure. What is the length of the resulting bar in meters?

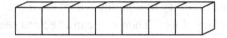

(S5K-13-8)

Problem 3.11:
A hollow rectangular parallelepiped measuring $6 \times 5 \times 4$ is composed of small $1 \times 1 \times 1$ cubes as shown. There are no holes in the surface of the shape, and there are no further cubes in the inside. How many of the cubes used to build the shape are not visible in the figure?

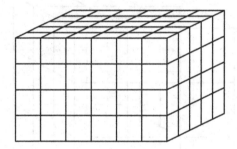

(S5K-15-3)

Problem 3.12:
Amar glues a red unit cube to each face of a white unit cube and ends up with a completely red object. He then glues black unit cubes to each face of the red object, with some of the black cubes being glued to two red faces. Determine the surface area of the resulting black object.

(S5K-16-4)

TIPS

Tip for Problem 3.1:
This problem doesn't require a whole lot of reasoning, just spatial imagination. In your mind, you will have to move the objects around in space.
You might find it useful to try to imagine where the object made up of four
cubes can go in the big object. Then you can check to see if there is room
for the smaller object left over.

Tip for Problem 3.2:
There are a number of different ways to count this. Either take a look at the
number of square faces showing on the outside or take a look at how many
square faces of each of the small cubes are glued to a face of another small
cube. Just make sure you don't miss any!

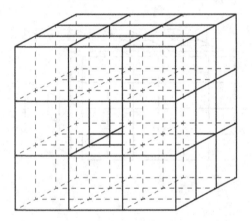

Tip for Problem 3.3:
The idea behind this problem is practically the same as in the previous
problem. If you can get that one, this one should be a snap!

Tip for Problem 3.4:
There are only four possible values for the number of neighbors a small
cube can have in the big cube, depending on the location of the small cube
relative to the corners and the edges. Also, the number of small cubes of
each of these four types depends on the value of n, if the big cube is built up
out of $n \times n \times n$ small cubes.

Tip for Problem 3.5:
If you find it difficult to imagine the connection between the number of bricks and the number of squares, try out what happens with smaller numbers! What if the rectangle measures 2 × 3 units, and the wall is only 2 units high? Can you see the connections between the numbers?

Tip for Problem 3.6:
Where does a small cube have to be placed on the big cube, if we know that exactly two of its faces are going to be painted green? We know that there are small cubes in the middle that aren't painted at all. We also know that corner cubes will have three sides painted, for instance.

Tip for Problem 3.7:
It will be helpful to use a variable. Name the height h. How many cubes will the entire block be made up of? How many of those will be red?

Tip for Problem 3.8:
Look at the right-hand (visible) side of the messed-up solid. There are two little rectangles visible here that don't exist in the properly glued solid. What does this mean for the left-hand (not visible) side of the solid?

Tip for Problem 3.9:
There are a number of square holes in the resulting object. Each of these holes was created by removing five of the little cubes. A good start is to count the number of holes and multiply by 5. Watch out, though! Some of these holes pass through the same spots!

Tip for Problem 3.10:
How many small cubes are there in each row of the big cube?

Tip for Problem 3.11:
It isn't hard to figure out how many cubes would be used to build such a figure if it were solid. Once we know that, we can figure out how many we need to fill the hole inside it, since the inside is just another, smaller, rectangular block. Subtraction will tell us how many cubes were used altogether, and then we are almost done.

Tip for Problem 3.12:
First of all, it will be helpful to figure out how many black cubes there are in the finished object. All of these are glued to at least one other cube, but some are glued to more than one face of another.

SOLUTIONS

Solution for Problem 3.1:

The one that isn't possible is (e).

You can see how the other four can be formed in this picture:

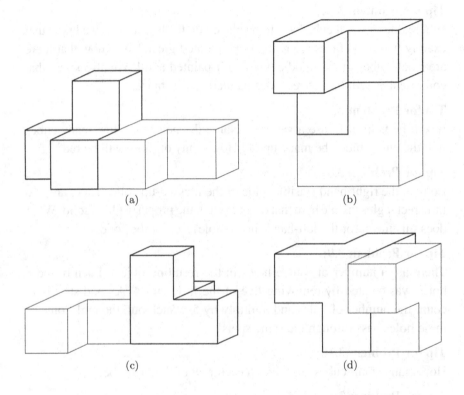

(a)   (b)

(c)   (d)

Just because we have a way to form these four, doesn't mean that (e) is impossible, of course. (On the other hand, if we assume that the question was posed in a way that made sense, it sort of does ...) In order to see why this really is the case, it is helpful to take a look at the "inside" cube of object (e). No matter how we place the original four-cube object inside this, the "inside" cube will be part of it, and either the cube on the left or the one on the right will be separated from the other remaining cubes. Either way, there is no way to build this out of the two original objects.

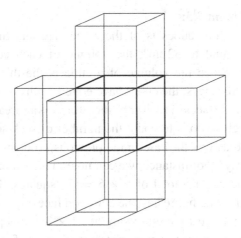

Solution for Problem 3.2:

One way to do this, is to count the number of small squares showing in each corner and on each edge of the big cube. There are eight small cubes placed in the corners of the big cube, and three of their faces are outside and have to be painted. Similarly, there is a small cube in the middle of each of the twelve edges of the big cube, and four square faces of each of these have to be painted. This means that the total number of small squares to be painted is equal to

$$8 \times 3 + 12 \times 4 = 72.$$

Another way to arrive at this number, is to calculate "backwards". The skeleton is made up of $8 + 12 = 20$ small cubes, each of which has 6 square faces, for a total of $20 \times 6 = 120$. (For instance, $3 \times 3 \times 3 = 27$ small cubes in the big cubes, out of which one from each of the six faces has been removed, along with the small cube in the middle of the big one.) Not all of these will be painted, however. Three square faces of each of the eight small corner cubes will not be painted, as they are glued to adjacent small cubes. The same is true for two square faces of each of the twelve small cubes in the centers of the edges of the big cube. This reasoning also leaves

$$120 - 8 \times 3 - 12 \times 2 = 72$$

small squares to be painted.

Solution for Problem 3.3:

Since each of the four cubes is of the same size and the total volume of the object is equal to $32\,\text{cm}^3$, the volume of each cube is equal to $32 \div 4 = 8\,\text{cm}^3$. Since the volume of a cube is the third power of the length of each of its edges, the cubes have edge lengths of $\sqrt[3]{8} = 2\,\text{cm}$.

Each of the square faces of each cube therefore has an area of $2^2 = 4\,\text{cm}^2$.

All we have left to do is to count the number of such squares showing on the object, and as was the case in the previous problem, we can do this in a number of ways. For instance, we can note that each of the four cubes has six square faces, for a total of $4 \times 6 = 24$ squares. The cube in the bottom back right of the figure has one common face with each of the other three cubes, and we must therefore subtract $2 \times 3 = 6$ squares from this total, since these common faces are not part of the surface of the object. This leaves us with $24 - 6 = 18$ exposed squares, and since each has an area of $4\,\text{cm}^2$, the total surface area of the object is equal to

$$18 \times 4\,\text{cm}^2 = 72\,\text{cm}^2.$$

Solution for Problem 3.4:

We already know that a small cube in the interior of the large cube has six neighbors, namely one on each side. In the figure below, we see what happens when we remove a small cube either in the center of a face, on an edge, or in a corner. We can see that these removals expose 5, 4 and 3 small square faces of the small cubes respectively, and these are the numbers of neighbors the small cubes of each type have in the large cube. Less than three neighbors is obviously not possible.

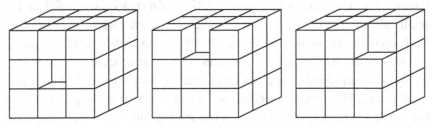

The small cubes with the second-largest number of neighbors are therefore the small cubes on the interior of the faces of the large cube (with 5 neighbors each), and those with the smallest number of neighbors are the ones in the corners (with 3 neighbors each).

The property described in the problem (i.e. the number of small cubes with the second-largest possible number of neighbors being three times as big as the number of small cubes with the smallest possible number of neighbors) therefore means that the number of interior face cubes is three times the number of corner cubes. If the large cube is made up of $n \times n \times n$ small cubes, there are $(n - 2)^2$ face cubes on each of the six faces, and there are certainly eight corner cubes. We can therefore write the property in the form

$$6 \times (n - 2)^2 = 3 \times 8.$$

Dividing this equation by 6 gives us $(n - 2)^2 = 4$, or $n - 2 = 2$, i.e. $n = 4$. The number of small cubes in the large cube is therefore equal to $4^3 = 64$.

We can check that this fulfills the required property by checking the number of small face cubes in a $4 \times 4 \times 4$ cube. There are $2^2 = 4$ face cubes on each of the six faces, and therefore a total of $6 \times 4 = 24$ altogether. This is indeed three times the number of corners, since $3 \times 8 = 24$ is certainly correct.

Solution for Problem 3.5:

Surrounding a rectangle made up of squares of the same size can be done by lengthening each side strip of the rectangle by one unit. Look at the picture below:

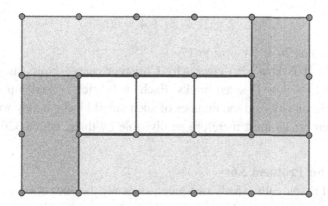

In the middle, we have a rectangle measuring 3×1 units. In order to build a wall around this rectangle, we add two light gray rectangles at the top and the bottom, each of which is $3 + 1 = 4$ units long, and two darker gray rectangles at the left and right, each of which is $1 + 1 = 2$ units long. The total number of unit squares required for the wall is therefore $2 \times 4 + 2 \times 2 = 12$.

This can be stated more generally. If the center rectangle measures $m \times n$, we add two rectangles at the top and the bottom, each of which is $m + 1$ units long, and two rectangles at the left and right, each of which is $n + 1$ units long. The total number of unit squares required for the wall is then $2(m+1)+2(n+1) = 2m+2n+4$. Another way to see this, is by considering the squares to be placed in the corners of the large rectangle surrounding the original one. Of course, there are four such corners. The strips above and below the original rectangle each have the length m, and the strips at the left and right each have the length n. Adding up the lengths of these strips and the four corners also results in the expression $2m + 2n + 4$.

Now, we are ready to consider the problem at hand. The original rectangle measures 14×16. The surrounding wall therefore requires $2 \times 14 + 2 \times 16 + 4 = 64$ cubic blocks in each layer, and since the wall is to be 5 cubes high all around, Christian will require $64 \times 5 = 320$ such blocks.

If he now wants to surround a $(21 \times n)$-sized rectangle with a wall four blocks high, the length of the wall will therefore be $320 \div 4 = 80$. Since we know a general expression for the length of such a wall, we obtain

$$2 \times 21 + 2n + 4 = 80,$$

which gives us $2n = 34$, or $n = 17$.

Finally, it is now also clear why Christian cannot build his original wall with $(3 \times 1 \times 1)$-sized bricks. Each such brick is made up of three small cubic blocks, and the number of such small blocks in any wall built out of such bricks must therefore be divisible by three, which 320 clearly is not.

Solution for Problem 3.6:
The small cubes with exactly two green sides are the ones that were on the edges of the large cube, but not directly in the corners. Three such small cubes have been removed in the right-hand figure on the next page, and it is easy to see why two of their sides will have been painted.

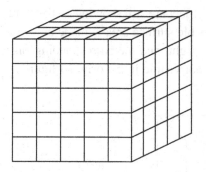 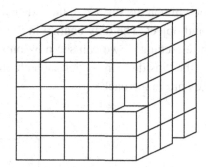

We therefore simply wish to find out how many such interior edge cubes there are. The big cube has 12 edges, and there are three interior edge cubes along each of these. That gives us a total of $12 \times 3 = 36$ small cubes with exactly two green sides.

Solution for Problem 3.7:
If we let h denote the height of the completed block, we can first note that the block will be made up of $8 \times 10 \times h = 80\,h$ cubes. Since the outside cubes in each direction are always red, the blue cubes are the ones not touching any outside face of the block. In each of the spatial directions, these are two less than the total number in that direction, as the two outermost are red. This means that the total number of blue cubes equals

$$(8 - 2) \times (10 - 2) \times (h - 2) = 48h - 96.$$

This is exactly half the total number of cubes used in building the block. From this, we get

$$2 \times (48h - 96) = 80h,$$

and this is equivalent to $96h - 192 = 80h$ or $16h = 192$, which yields $h = 12$.

As usual, we can check if this result fulfills the properties of the problem. Indeed, if the height of the block is 12, the block is composed of a total of $8 \times 10 \times 12 = 960$ cubes. The number of blue cubes is equal to $6 \times 8 \times 10 = 480$, which is exactly half the total number of cubes in the block, as required.

Solution for Problem 3.8:
If we think of the object on the right resulting from the object on the left by pushing the middle cube to the right, we can see that all the squares

showing on the surface of the left-hand object are still showing on the right-hand object. The only change in the visible surface area is made up of the little rectangles we can see now on the right, and the corresponding ones on the "inside", on the left (which are not visible in the original figure, as they are in the back).

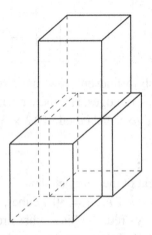

Altogether, there are four such rectangles, two on the right and two on the left, and each of these has an area of $1\,\text{cm} \times 6\,\text{cm} = 6\,\text{cm}^2$, as the cube with edge length $6\,\text{cm}$ has been pushed out by $1\,\text{cm}$. The total added area is therefore equal to $4 \times 6\,\text{cm}^2 = 24\,\text{cm}^2$.

Solution for Problem 3.9:
We can count the square holes poking through the big cube. There are four such holes in each of the three directions in space (front to back, left to right, top to bottom), for a total of $3 \times 4 = 12$ such holes. Each of these consists of three small cubes only belonging to that specific hole and two in the interior, that it has in common with two other holes going in the other two directions. Altogether, there are $2 \times 2 \times 2 = 8$ such small cubes in the interior, each belonging to three of the holes. This means that the total number of small cubes that have been removed is equal to the sum of these 8 and the $12 \times 3 = 36$ of the small cubes belonging exclusively to on of the square holes, making a total of $8 + 36 = 44$ small cubes that have been removed. Since the original big cube was made up of $5 \times 5 \times 5 = 125$ small cubes, the number left after making the holes is equal to $125 - 44 = 81$.

Solution for Problem 3.10:

Since $5 \times 20 = 100$, there are 20 small cubes with sides of length 5 cm in each row of the big cube with sides of length 100 cm $= 1$ m. The big cube is therefore made up of $20 \times 20 \times 20 = 8000$ small cubes. Each of these has sides of length 5 cm, and the total length of the bar is therefore equal to 8000×5 cm $= 40000$ cm $= 400$ m. Quite amazing when you think about it; 400 times as wide as the original cube!

Solution for Problem 3.11:

If the parallelepiped were solid, it would be made up of $6 \times 5 \times 4 = 120$ small cubes. Since there are two on the outside surfaces in any direction, the number of small cubes completely in the interior would then be $(6 - 2)(5 - 2)(4 - 2) = 4 \times 3 \times 2 = 24$, which means that the hollow object we are considering is made up out of a total of $120 - 24 = 96$ small cubes.

Of these, the number of visible cubes includes all $6 \times 5 = 30$ on the top surface, the $(6 \times 4) - 6 = 18$ in the front surface that are not also part of the top surface, and the $(5 \times 4) - 5 - 3 = 12$ on the right surface that are neither part of the top nor part of the front. Altogether, there are therefore $30 + 18 + 12 = 60$ visible small cubes. This leaves $96 - 60 = 36$ such small cubes, that are not visible in the figure.

Solution for Problem 3.12:

After the first step, Amar has an object with six red cubes showing, and a hidden white cube in the middle, as shown on the left. In the next step, he glues black cubes onto the showing faces of these red cubes. The black cubes that he glues onto the six red faces parallel to the faces of the white cube will have no common faces with any of the others, and will therefore each have 5 black faces showing, for a total of $6 \times 5 = 30$.

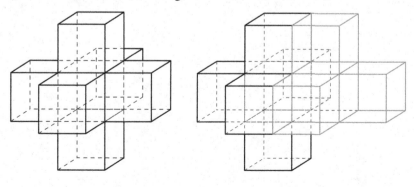

In the right-hand figure, we have added three of the black cubes, one on the far right and two that each have a common face with two of the red cubes. The number of such black cubes is equal to the number of edges of the original white cube, i.e. twelve, as there is one wedged between any two neighboring red cubes. Four faces of each of these are showing, as two are glued to red squares, giving us a total of $12 \times 4 = 48$ such faces showing. This means that a total of $30 + 48 = 78$ black faces are showing, and since each has the area 1 (the area of a face of a unit cube), the total surface area of the black object is equal to 78.

Chapter 4

Sports

Sports are fun and solving problems is fun, so solving problems about sports must be at least twice as much fun, right? Or is it fun squared?

Some readers might take issue with some of the activities classified as "sports" here (chess? folk dancing?), but try to keep an open mind. We start off with a number of problems concerning races: foot races, bicycle races, and because the problems were developed in Austria, ski races. Then we will advance to some games played in a circle, and then finally to some problems concerning more classical team sports.

PROBLEMS

Problem 4.1:
Four friends compare their times after a 60 m footrace. Anna was slower than Bianca and Christopher. Christopher wasn't as fast as Bianca, but faster than Dieter. Dieter's and Bianca's times have the same sum as those of Anna and Christopher. Which of them was the fastest and which was the slowest?

(S5K-98-2)

Problem 4.2:
Three girls compete in a 2000 m race. Maria wins, 200 m ahead of Anne and 290 m ahead of Katharina. How much ahead of Katharina will Anna be when she reaches the finish line if both of them continue at the same pace?

(S5K-01-8)

Problem 4.3:
Three athletes start a 100 m race simultaneously. As the first athlete reaches the finish line, the second still has 10 m to go. As the second reaches the finish line, the third still has 10 m to go. How far was the third athlete from the finish line when the first athlete crossed it?

(StU-93-10)

Problem 4.4:
Stefan and Christoph are running in the woods with their dog Rex. Stefan completes the whole 3 km course in 15 minutes. It takes Christoph 20 minutes to do the same. The whole time Christoph is running, Rex runs back and forth between them at a pace that is twice as fast as Stefan's. How far does Rex run altogether?

(S5K-00-4)

Problem 4.5:
Niklas and Paula run several laps around a lake. They start together at the hollow tree and run until they arrive simultaneously at that spot again for the first time. Paula runs a lap around the lake in 12 minutes, but it only takes Niklas 10 minutes. How many minutes will they run until they stop?

(S5K-17-1)

Problem 4.6:
Fred is training for a bicycle race. He plans to ride 8 km more on each of his training days than he did the day before. After 11 days of training, he adds his distances and finds that he has completed a distance of 1045 km altogether. How far did he ride on the fourth day?

(S5K-02-3)

Problem 4.7:
After the first heat of the club championships of the Hinterholz ski club, Sepp Einfädler was in seventh place and his brother Karl was eighth. In the second heat, Sepp had the eighth-best time and won the championship on aggregate times. Karl had the seventh-best time in the second heat and came in last in aggregate times. All competitors were classified. How many skiers took part in the championship?

(StU-02-A9)

Problem 4.8:
In the World Cup of alpine skiing, the winner of a race is awarded 100 points, the second place finisher is awarded 80 points and the third place finisher is awarded 60 points. After nine races, Marcel Hirscher had 800 points. We know that he finished every race among the top three.

(a) How many races did he win at least?
(b) How many races did he win at most?
(c) How many third place finishes did he have at most?

(StU-13-B6)

Problem 4.9:
At a course in folk dancing, all the participants are standing in a big circle, and they are counted off. Dancer number 20 is standing right across from dancer number 53. How many dancers are there in the circle?

(O-93-7)

Problem 4.10:
Today, the folk dancing group is practising a slap dance, in which all the members are, as usual, arranged in a big circle. During the course of the dance, each dancer gets slapped 5 times by each of his or her direct neighbors and twice by their neighbors' neighbors in the circle. (They do not slap themselves, however.) They also get slapped once by every other dancer. After the dance is over, each dancer has been slapped 16 times. How many dancers are there in the circle today?

(O-96-1)

Problem 4.11:
Six friends, Alf, Beate, Chris, Doris, Edgar and Fritz are standing in a circle to play a ball game. Beate and Doris want to stand next to each other, as do Edgar and Fritz. Unfortunately, there has been a misunderstanding in the group, and neither Edgar nor Fritz is willing to stand beside Beate or Doris. How many different arrangements are possible for the six players? (We say that two arrangements are the same, if every one of the friends is standing between the same two people in each of them.)

(StU-14-A6)

Problem 4.12:
Alice, Harald and Mitzi are organizing a sports day with various competitions. All three of them take part in each competition. The points they are awarded for a first, second or third place finish in each competition are all positive integers. There are no ties. A first place result is awarded more points than a second place, and this in turn is awarded more points than a third place. At the end of the sports day, Alice has 18 points, Mitzi has 9 and Harald has 8, even though he won the sack race. How do we know that there were exactly 5 different competitions? How many points were awarded for a first, second or third place finish? Who won the somersault competition?

(StU-00-B7)

Problem 4.13:
Some 13- and 14-year-olds are playing a chess tournament. Every participant plays one game against every other one. In 15 of the games, two 14-year-olds play each other, and in 6 of the games, two 13-year-olds play each other. How many games are played, in which a 13-year-old plays against a 14-year-old?

(StU-03-A9)

Problem 4.14:
Five teams are competing in a tournament. In the course of the tournament, each team plays every other one exactly once. Each team gets 2 points for a win, 1 point for a draw and 0 points for a loss. At the end of the tournament, the winning team has 6 points, the second-place team has 5 points, and the three other teams are all tied with the same number of points. How many points does each of these teams have?

(O-95-2)

Problem 4.15:
Seven soccer teams from the little town of Semriach are competing in their village league for the local championship. Every team plays every other team twice in a season. How many games are played in a season of the league altogether?

(S5K-03-3)

Problem 4.16:
The sports club *Les Miserables* has two teams, one for gymnastics and one for hockey. The 12 gymnasts have an average weight of 81 kg. The 18 hockey players have an average weight of 46 kg. What is the average weight of all the club members?

(S5K-98-4)

Problem 4.17:
At a sports academy with 300 students, 60% of the students play volleyball in the summer, while the other 40% play basketball. In the winter, each student is either a skier or a skater, but no one is both. 56% of the skiers play volleyball in the summer and 30% of the volleyball players skate in the winter. How many of the students are basketball playing skaters?

(O-93-6)

Problem 4.18:
A sports club has three sections: football, tennis and bowling. Each club member belongs to exactly one section. The soccer section has 23 members, the tennis section has 12 members. The remaining 30% of the club members are bowlers. How many members does the sports club have?

Another sports club with the same three sections has 25 members that play soccer, and 17 members that play tennis. Among them, there are several members that belong to both the football section and the tennis section. One-quarter of all club members are bowlers. None of these plays soccer or tennis. What is the smallest possible number of members this sports club could have under these circumstances?

(StU-08-B1)

Problem 4.19:
The football teams from North Sturm City (NSC) and South Sturm City (SSC) are traditional rivals. They played each other twice this year, and scored a total of eleven goals in the course of these matches. Karl claims that the first match ended in a draw, and that SSC scored three times as many goals as NSC in the second match. Explain why Karl must be mistaken.

Daniel correctly states that the first match did indeed end in a draw, but that SSC only scored twice as many goals as NSC in the second match.

In addition, he notes that more goals were scored in the second match than the first. Determine the results of both matches.

(StU-08-B3)

Problem 4.20:
In basketball, a successful basket attempt can be worth 1, 2 or 3 points. The Graz Ballesterers scored 18 times in the first quarter. They scored more 2-point shots than 1- and 3-point shots put together. In doing this, they scored at most H points, but at least L. Determine H − L.

(StU-14-A9)

TIPS

Tip for Problem 4.1:
You can figure out in which order Anna, Bianca and Christopher finished by looking at the first two bits of information. Then, just take a look at the information you haven't used yet.

Tip for Problem 4.2:
How far has Anne run when Maria crosses the finish line? How much farther does she have to go? How much has she already gained on Katharina so far? How much more will she gain by the time she finishes?

Tip for Problem 4.3:
This one is so similar to Problem 2, right?

Tip for Problem 4.4:
This is easier than it might seem at first glance. If you read the text carefully, you will note that you know how long Rex runs, and also how fast he is running.

Tip for Problem 4.5:
This is actually a lot like Problems 2 and 3, if you think about it. Niklas is faster than Paula, and once again it will be helpful to express her speed as a fraction of his.

Tip for Problem 4.6:
Try writing this one like a standard addition, using a variable for something you would like to know the value of, but don't yet. You can actually guess

the right answer if you think about this one carefully. Of course, you will then have to check, whether your guess was right.

Tip for Problem 4.7:
Somehow this doesn't seem possible at all, right? It is, though! Since Sepp came in eighth in the first heat, there must have been at least eight competitors. Is it possible that there were exactly eight? Could there have been twice that many? Why aren't either of these options possible?

Tip for Problem 4.8:
Marcel could have earned 900 points in those nine races, but he only earned 800. This means that he failed to win exactly 100. We know that he did earn either 100, 80 or 60 points in each race, and that he failed to earn either 20 or 40 points in some of them. Some combination of these must therefore add up to the 100 points he failed to earn in total.

Tip for Problem 4.9:
Think of a smaller circle. What numbers are across from each other in a circle with just four members? How about a circle with six members? Who is across from number 1?

Tip for Problem 4.10:
Once you know the total number of slaps each dancer gets from the two people on either side, the rest should come quite easily.

Tip for Problem 4.11:
Note that there are two couples that each definitely want to stand together. You can essentially treat each of these couples as one person.

Tip for Problem 4.12:
Check out how many points were awarded altogether. What does this total tell you about the number of points awarded for each individual competition?

Tip for Problem 4.13:
How many 14-year-olds were there in the tournament? How many 13-year-olds?

Tip for Problem 4.14:
If there are five teams, how many games are played altogether? What does that tell us about the total number of points the teams had at the end?

Tip for Problem 4.15:
Imagine that every team has a home field. How many games are played on each of these fields in the course of a season?

Tip for Problem 4.16:
You probably shouldn't worry too much about the fact that the gymnasts seem to be a lot heavier that the hockey players. Maybe the hockey players are all school kids, while the gymnasts are all adult lumberjacks? Anyway, try figuring out the total weight of all the club members. That should get you there.

Tip for Problem 4.17:
There is a lot of information given here, but the order its given in might seem a bit confusing. You might consider thinking about this problem backwards.

Tip for Problem 4.18:
Just keep thinking about opposites. If 30% of the group are one thing, we know that 70% aren't, for instance.

Tip for Problem 4.19:
If a football match ends in a draw, it should be obvious that the number of goals scored in the match must be even.

Tip for Problem 4.20:
How many 2-point shots could they have scored?

SOLUTIONS

Solution for Problem 4.1:
Since Anna (A) was slower than Bianca (B) and Christopher (C) and Christopher wasn't as fast as Bianca, these three came finished in the following order A – C – B (from slowest to fastest). Christopher was also faster than Dieter (D), and so the only two possible orders for all four are D – A – C – B and A – D – C – B, but if it were the second of these, and Dieter was faster than Anna, which would mean that Dieter's and Bianca's times together would be better than those of Anna and Christopher. Their only possible finishing order is therefore D – A – C – B, and we see that Bianca was the fastest and Dieter was the slowest.

Solution for Problem 4.2:
When Maria crosses the finish line, Anne has run $2000 - 200 = 1800\,\text{m}$, and has another $200\,\text{m}$ to go. That is exactly $\frac{200}{1800} = \frac{1}{9}$ of the distance she has already run. So far, she has gained $290 - 200 = 90\,\text{m}$ on Katharina, and if they both run one-ninth of the distance they have run so far at the same pace, she will gain one-ninth of that again. In other words, in addition to the $90\,\text{m}$ she is already ahead of Katharina at that point, when she gets to the finish line, she will be another $\frac{1}{9} \times 90 = 10\,\text{m}$ ahead for a total lead of $90 + 10 = 100\,\text{m}$ at the finish line.

Solution for Problem 4.3:
Since the second athlete has only run $100 - 10 = 90\,\text{m}$ by the time the first has finished, he runs $\frac{90}{100} = \frac{9}{10}$ as fast the first. Similarly, the third athlete has only run $100 - 10 = 90\,\text{m}$ by the time the second has finished, and he therefore runs $\frac{90}{100} = \frac{9}{10}$ as fast the second. This means that the third athlete runs $\frac{9}{10} \times \frac{9}{10} = \frac{81}{100}$ as fast as the first. This means that he only ran $81\,\text{m}$ in the time the first athlete took to run the full $100\,\text{m}$, and he was therefore $100 - 81 = 19\,\text{m}$ from the finish line, when the first athlete crossed it.

Solution for Problem 4.4:
Since Rex is running for the same length of time as Christoph, he is running for 20 minutes, or one-third of an hour. Also, since he is running twice as fast as Stefan, and Stefan is running at a speed of $3\,\text{km}$ per quarter hour (or $3 \times 4 = 12\,\text{km}$ per hour), he is running at a speed of $24\,\text{km}$ per hour. In one-third of an hour, he therefore runs a total distance of $24 \div 3 = 8\,\text{km}$.

Solution for Problem 4.5:
Since Paula takes 12 minutes to run the same distance it takes Niklas 10 minutes to run, her speed is $\frac{10}{12} = \frac{5}{6}$ of his. In other words, taking a close look at the numerator and denominator of this fraction; Paula will run 5 laps in the time it takes Niklas to run 6 laps of the lake. Since they have to each run a whole number of laps to return to the hollow tree, and the difference in these numbers is only $6 - 5 = 1$, this is the smallest number of laps each of them must run to meet again at the hollow tree. It takes Niklas $10 \times 6 = 60$ minutes (and Paula $12 \times 5 = 60$ minutes) to do this.

Solution for Problem 4.6:
We aren't told how many km Fred rides on the first day of his training program, but everything else depends on that one number. It therefore seems

like a reasonable idea to introduce a variable x as the number of km Fred rides that first day. This means that he rides $x + 8$ km the second day, $x + 2 \times 8$ km the third day, $x + 3 \times 8$ km the fourth, and so on, until he eventually rides $x + 10 \times 8$ km on the eleventh day of his training. Altogether, this means that he rides a total distance of

$$x + (x + 8) + (x + 2 \times 8) + (x + 3 \times 8) + \cdots + (x + 10 \times 8) \text{ km},$$

and since we are given that this total distance is equal to 1045 km, we can write

$$x + (x + 8) + (x + 2 \times 8) + (x + 3 \times 8) + \cdots + (x + 10 \times 8)$$
$$= 11x + (8 + 2 \times 8 + 3 \times 8 + \cdots + 10 \times 8) = 1045.$$

In order to calculate the value of the expression in the brackets, we get

$$8 + 2 \times 8 + 3 \times 8 + \cdots + 10 \times 8 = 8 \times (1 + 2 + 3 + \cdots + 10)$$
$$= 8 \times 55 = 440,$$

and this gives us the equation $11x + 440 = 1045$. This is equivalent to $11x = 605$ or $x = 55$, and we see that Fred rode 55 km on his first day of training, and therefore $55 + 3 \times 8 = 55 + 24 = 79$ km on his fourth day.

Solution for Problem 4.7:

Since Sepp won on aggregate, he must have had a better time in the second heat than any of the six competitors ahead of him in the first. This means that the six who placed better than Karl (and Sepp) in the second heat must all have been placed behind Sepp in the first. Since there were therefore six competitors placed better than the two brothers in each heat, and none of these was placed better in both, there must have been at least $6 + 6 = 12$ competitors in the race in addition to the two brothers, for a total of at least 14. On the other hand, there cannot be a competitor who finished behind Karl in both heats, as he would then have an aggregate time worse than Karl, and it would follow that Karl could not place last. This means that every participant in the race placed ahead of Karl in one heat, and it follows that there cannot have been more than the 14 participants already identified.

To see that 14 is indeed a possible number, let us consider the following table.

Name	Time heat 1	Rank heat 1	Time heat 2	Rank heat 2	Total time	Rank
Sepp	30″	7	33″	8	63″	1
Racer 1	22″	1	43″	9	65″	2
Racer 2	23″	2	44″	10	67″	3
Racer 3	24″	3	45″	11	69″	4
Racer 4	25″	4	46″	12	71″	5
Racer 5	62″	9	11″	1	73″	6
Racer 6	26″	5	48″	13	74″	7
Racer 7	64″	10	12″	2	76″	8
Racer 8	28″	6	49″	14	77″	9
Racer 9	66″	11	13″	3	79″	10
Racer 10	68″	12	14″	4	82″	11
Racer 11	70″	13	15″	5	85″	12
Racer 12	72″	14	16″	6	88″	13
Karl	60″	8	32″	7	92″	14

Note that the numbers here were chosen specifically in order to avoid ties, but there are many ways to create such tables for 14 participants.

Solution for Problem 4.8:
We will say that Marcel "lost" points if he failed to earn them. In other words, if he came in second in some race and earned 80 points (and not 100, as he would have for a win), we say he lost 20 points in that race. Similarly, we say that he lost 40 points for a third place finish (since he earned only 60 instead of the possible 100 for a win).

(a) We know that he lost 100 points over the course of the nine races. Since he cannot lose less than 20 points in an individual race, there can be at most 5 races in which he did not win, as $100 = 5 \times 20$. He therefore won at least $9 - 5 = 4$ races.

(b) Similarly, he cannot lose more than 40 points in an individual race. There can therefore not be less than three races in which he did not win, as $100 = 2 \times 40 + 20$, but $100 < 3 \times 40$. He therefore won at most $9 - 3 = 6$ races.

(c) The argument from part (b) also answers this question, as three or more third place finishes would have meant him losing at least 120 points. He therefore had at most two-third place finishes.

Solution for Problem 4.9:
If number 20 is standing across from number 53, his neighbor, number 19, is across from number 53's neighbor, namely 52. Next, this means that 18 is across from 51. Continuing this reasoning, we can see that number $1 = 20 - 19$ is across from number $53 - 19 = 34$. One half of the circle is therefore numbered from 1 to 33, with the other half starting at 34, and this second half must therefore finish with number $2 \times 33 = 66$.

If you find this difficult to visualize, consider the following figure.

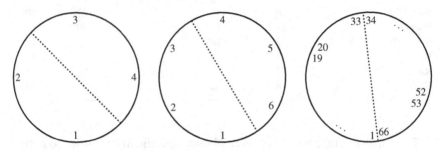

In the left circle, the participants are numbered from 1 to 4. The dashed line separates the first half (1 and 2) from the second half (3 and 4), and the first number in the second half (3) is across from number 1. In the middle circle, we have the analogous situation for 6 participants. Once again, the first half (1 to 3) is separated from the second (4 to 6) by a dashed line, and the first number of the second half (4) is across from number 1. Finally, on the right, we have the situation of the problem. Since 20 is across from 53, 1 is across from 34. This means that 33 is the last number in the first half (with the two halves once again separated by a dashed line), and this means that there are 66 participants in the circle altogether.

Solution for Problem 4.10:
Since each of the dancers gets slapped five times by each neighbor and twice by each of their respective neighbors, they get slapped a total of $2 \times 5 + 2 \times 2 = 14$ times by these four people. Since they get slapped 16 times altogether, there are two slaps missing in this calculation, and these are doled out by two further dancers in the circle, as we know that they get slapped by each of the other dancers once. This means that there are two more people in addition to the person getting

slapped and the four neighbors, for a total of seven dancers in the circle altogether.

Solution for Problem 4.11:
Well, at least they aren't slapping each other this time, but they still aren't playing nice, are they? Who knows why some people refuse to stand beside some others?

Since Beate and Doris want to stand together, we can treat them as one person, that we will call B/D, and the same holds for Edgar and Fritz, who we shall call E/F. Alf (A) and Chris (C) have to be placed between these pairs, and this means that we have to place them in the order

$$A - B/D - C - E/F.$$

By choosing which of each of the members of the two pairs will stand next to Alf, then uniquely determines the arrangement. Since there are two options for each pair (Alf can stand either next to Beate or Doris and next to either Edgar or Fritz), there exist a total of $2 \times 2 = 4$ different arrangements possible under these conditions.

Solution for Problem 4.12:
The total number of points awarded at the sports day is equal to $18 + 9 + 8 = 35$. We know that there was more than one competition, since there was both a sack race and a somersault competition. We also know that at least six points were awarded for each of the competitions, since the smallest possible point scores under the given conditions are 1 for a third place finish, 2 for a second and 3 for a first, for a total of six. Noting that $35 = 5 \times 7$ is the product of two primes, this means that there must have been 5 different competitions, with 7 points awarded at each competition. The only way that 7 points can be awarded in an ascending manner for three prizes is 1 point to be awarded for a third place finish, 2 for a second and 4 for a first.

Harald won the sack race, and since he got 4 points for that first place, he must have come in third in all four other competitions, since he got at least one point in each of them, but only a total of $8 - 4 = 4$ points for all four. This means that Mitzi can have at most 1 third place finish, and must have received at least 2 points in each of the other four competitions, for a total of $1 + 4 \times 2 = 9$. Since this smallest possible number of points she can have achieved is indeed her total score, she did not win any of the competitions, and it follows that Alice won all the other four competitions, including the somersault competition.

Solution for Problem 4.13:

Let us assume that there were n 14-year-olds in the tournament. Each of these plays against each of the $n - 1$ others. This means that the total number of games played among them is exactly half of $n \times (n - 1)$, since we would count every game played twice if we just multiply the number of players by the number of their opponents. Since this number is 15, we have to find two successive integers, whose product is equal to $2 \times 15 = 30$. These numbers are 5 and 6, and we therefore know that there were six 14-year-olds in the tournament. Similarly, since the 13-year olds played six games, we obtain the number of 13-year-olds in the tournament by expressing $2 \times 6 = 12$ as the product of two successive integers. Since $3 \times 4 = 12$, there were four 13-year-olds in the tournament. We know that each of the six 14-year-olds played each of the four 13-year-olds once, and the total number of such games was therefore equal to $6 \times 4 = 24$.

Solution for Problem 4.14:

If there are five teams, the total number of games played altogether is equal to half of 5×4, or 10. This tells us that a total of 20 points are awarded in the course of the tournament, since there are always two points awarded in each game, no matter what the outcome is. Since the first two teams got $6 + 5 = 11$ points, that leaves $20 - 11 = 9$ for the other three teams, and since they are all tied, they each got $9 \div 3 = 3$ points.

While this solves the problem, it is worthwhile to give some thought as to whether this result is actually possible. One way this could happen is for the winning team to beat all three of the teams tied for third and lose to the second-place team, with all other games ending in a tie.

Solution for Problem 4.15:

If we imagine that each team plays every other team twice, we can interpret this as meaning that they all have a home game and an away game against every other team in the course of a season. This means that each of the seven teams will play 6 home games, for a total of $7 \times 6 = 42$ games in total played on all the fields.

Solution for Problem 4.16:
Since the total weight of all the gymnasts is equal to $12 \times 81 = 972$ kg and the total weight of the hockey players is equal to $18 \times 46 = 828$ km, the total weight of all $12 + 18 = 30$ club members is $972 + 828 = 1800$ kg. This means that their average weight is equal to $1800 \div 30 = 60$ kg.

Solution for Problem 4.17:
We know that 60% of the 300 students, or 180 students, play volleyball, and that therefore $300 - 180 = 120$ of them play basketball. Since 30% of the volleyball players also skate, we know that there are 30% of 180, or 54 students, who participate in these two sports. This leaves $180 - 54 = 126$ volleyball players who also ski. We know that this group makes up 56% of all the skiers, and there are therefore $\frac{126 \times 100}{56} = 225$ skiers altogether. This leaves a total of $300 - 225 = 75$ skaters, and we know that 54 of these play volleyball. This leaves a total of $75 - 54 = 21$ basketball playing skaters.

Solution for Problem 4.18:
Since the $23 + 12 = 35$ members of the sports club that aren't bowlers make up the other 70% of the membership, the total number of members in the sports club is equal to $\frac{35 \times 100}{70} = 50$.

Moving on to the second question, we know that one-quarter of the club members are exclusively bowlers, and this means that three-quarters of the club members aren't. The number of non-bowlers must therefore be divisible by 3. There are at least 25 non-bowlers, since there are 25 soccer players, but 25 is not divisible by 3. The smallest number divisible by 3 and also larger than 25 is 27. This implies that one-third of this number is the number of bowlers, and there are therefore 9 bowlers in the sports club, in addition to the 27 non-bowlers, for a total of $27 + 9 = 36$ club members. (Note that this means that there are 15 tennis players that also play football and 2 that don't.)

Solution for Problem 4.19:
If Karl were right, the number of goals scored in both matches would be even. A draw in the first match certainly means an even number of goals were scored there. If the winning team of the second match scored three times as many goals as the other team, the total number of goals scored by both teams in the second match would have been four times the number of goals scored by the losing team alone, and this would certainly be an even

number. This means that Karl being right would imply that the total number of goals in both matches must also be even, which contradicts the fact that eleven goals were scored. Karl is therefore definitely mistaken.

Since Daniel knows that SSC scored twice as many goals as NSC in the second match, we know that NSC scored exactly one-third of the goals in that match. This means that the number of goals scored in that match must be divisible by three. Now, we can check all possible results of the first match, which we know to be a tie. If the first match ended 0:0, there would have been 11 goals scored in the second match, but 11 is not divisible by 3. If the first match ended 1:1, there were $11 - 2 = 9$ goals scored in the second, and since NSC scored one-third of them, the result of the second match would have been 6:3. This is certainly a result consistent with all the information we have, since there were 9 goals scored in the second match, and only 2 in the first. A 2:2 score in the first implies $11 - 4 = 7$ goals in the second, which is again not divisible by 3. A 3:3 score in the first implies $11 - 6 = 5$ goals in the second, which is also not divisible by 3. A 4:4 score in the first leaves $11 - 8 = 3$ goals for the second, but 3 is less than 8. Finally, a 5:5 score in the first leaves just $11 - 10 = 1$ goal for the second, which doesn't work on any level. We see that the first match was a 1:1 tie, and the second a 6:3 romp for SSC.

Solution for Problem 4.20:
Since they scored more 2-point shots than everything else put together, and they scored a total of 18 times, at least 10 of these scoring shots must have been for 2 points. The smallest possible number of points will be scored with as many 1-point shots as possible along with the smallest possible number of anything else. This means that, in this case, there would be $18 - 10 = 8$ 1-point shots and 10 2-point shots for a total of $L = 10 \times 2 + 8 \times 1 = 28$. Similarly, the largest possible number of points will be scored with as many 3-point shots as possible along with the smallest possible number of anything else. This means that, in this case, there would be $18 - 10 = 8$ 3-point shots and 10 2-point shots for a total of $H = 10 \times 2 + 8 \times 3 = 44$. In summary, this gives us $H - L = 44 - 28 = 16$.

Chapter 5

Circles

There are several interesting types of problems that deal with circles in some way. Such problems can have some aspect of counting, or possibly deal primarily with such fundamental geometric properties as distance, angle or area. Circles are such basic objects that all such problems end up with a common flavor, however.

PROBLEMS

Problem 5.1:
In the figure we see a circle, a square and a triangle placed in such a way that a total of 7 points of intersection result. What is the largest number of points of intersection that can result by placing a circle, a square and a triangle in an appropriate manner?

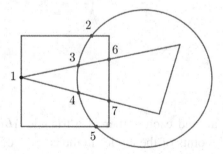

(S5K-11-5)

Problem 5.2:
Two circles are externally tangent, i.e. touch each other from the outside. One has a radius of 3 cm and the other has a radius of 5 cm. Determine the radius of the smallest circle that contains both circles completely.

(S5K-03-2)

Problem 5.3:
We are given a circle k with midpoint M and radius $r = 5$ cm and a point P set at a distance of 2 cm from M. A further point Q is placed 7 cm from P. Where can Q lie with respect to k? In other words, can Q lie inside k? Can Q lie on k? Can Q lie outside k?

(StU-00-A9*)

Problem 5.4:
In the figure, which is not drawn to scale, we see a triangle ABC with circumference 26 cm. The radius r_A of the circle k_A (with center A) is equal to 5 cm. How long is the side BC if the three circles are mutually tangent, as shown?

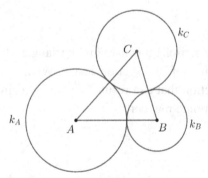

(S5K-08-3)

Problem 5.5:
A circle is drawn around each corner of a triangle ABC with perimeter 34 cm, with the midpoints of the circles in the corners of the triangle. The circles around A and B are externally tangent, as are the circles around B and C and the circles around C and A. The circle around A has the radius 5 cm and the circle around B has the radius 8 cm. Determine the radius of the circle around C.

(StU-02-A10B)

Problem 5.6:
A circle is drawn around each corner of a triangle ABC with sides of length $BC = a = 12$ cm, $CA = b = 15$ cm and $AB = c = 13$ cm, with the midpoints of the circles in the corners of the triangle. The circles around A and B are externally tangent, as are the circles around B and C and the circles around C and A. Determine the radius of the circle around C.

(StU-02-A10A)

Problem 5.7:
Two intersecting circles k_1 and k_2 with common radius $R = 32$ cm have midpoints M_1 and M_2, respectively. The points in which the circles intersect the line segment $M_1 M_2$ divide $M_1 M_2$ into three parts of equal length. Determine the radius of the circle internally tangent to both given circles k_1 and k_2 and simultaneously tangent to $M_1 M_2$.

(StU-98-B7)

Problem 5.8:
The line t_1 is tangent to the circle k_1 in A and to the circle k_2 in B. Furthermore, the line t_2 is tangent to the circle k_1 in C and to the circle k_2 in D, as shown. Furthermore, the line t_3 is also tangent to both the circle k_1 and the circle k_2 and intersects t_1 in E. How long is the line segment AB, if we know that $AE = 3$ cm and $CD = 7$ cm?

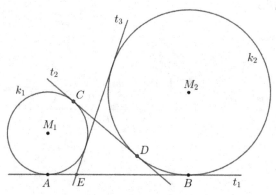

(StU-05-A9)

Problem 5.9:
A semicircle k_1 with center M is drawn on the segment AB as its diameter, as shown. On the other side of AB, a semicircle k_2 is drawn with the segment

AM as its diameter. Points C and D lie on k_1 and k_2, respectively, in such a way that CD passes through M.

(a) If we are given $AB = 10\,\text{cm}$ and $AD = 4\,\text{cm}$, determine the lengths of AC and BC.

(b) Determine β if we are given $\varepsilon = 45°$. How must ε be chosen, such that α and ε are the same size?

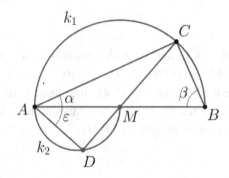

(StU-03-B7)

Problem 5.10:
In the given figure, M is the midpoint of circle k (with radius r). We are given $\alpha = 70°$. The points A, M and P lie on a common line, as do the points B, C and P. The distance from C to P is equal to r. Determine the angle $\beta = \angle APB$.

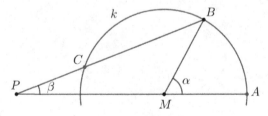

(S5K-05-9)

Problem 5.11:

In the given figure, which is not drawn to scale, we see a quadrilateral $ABCD$ and a circle with center C. The circle is tangent to AB in B and to AD in D. The angle γ is five times as big as the angle α. Determine the size of the angle α.

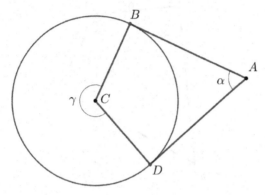

(S5K-14-6)

Problem 5.12:

In the isosceles trapezoid $ABCD$ (with AB parallel to CD), the midpoint M of AB is the center of the circumcircle. P lies on the (shorter) arc CD in such a way that $CP\|MD$ holds. The angles $\angle MDC$ and $\angle CMP$ are equal. Determine their measure.

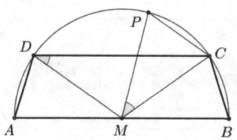

(StU-14-B6)

Problem 5.13:

Four points A, B, C and C' lie on a common circle with the diameter AB, as shown. C and C' are symmetric with respect to AB, and $BC = CC'$ holds. Determine the angle $\angle BAC$.

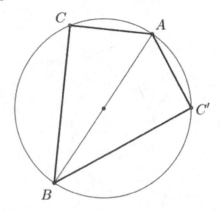

(StU-15-A5)

Problem 5.14:

Both the midpoints M_1 and M_2 of the circles k_1 and k_2 and their common points P and Q lie on the circle k. Determine the area of the gray region if the radii of the two circles k_1 and k_2 are both equal to $2\,\mathrm{cm}$.

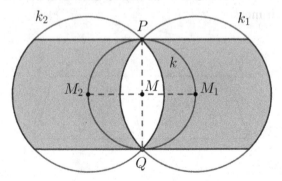

(S5K-05-7)

Problem 5.15:

A semicircle is drawn on a segment AB with length 200, and a chord AC of the semicircle is drawn from A with $AC = 160$. The line perpendicular to AB is drawn from C, and this line intersects AB in D. Furthermore, the line perpendicular to *AB* is drawn through the midpoint *M* of *AB*, and this line intersects *AC* in *E*. Determine the area of the quadrilateral *MDCE*.

(O-92-5)

Problem 5.16:
A circular table has the midpoint *M*. Points *A* and *B* are marked on the edge of the table in such a way that *AM* is perpendicular to *BM*. A circular tray is put on the table in such a way that its edge lies directly over the points *A*, *B* and *M* (with the tray reaching partly beyond the edge of the table). What fraction of the table's surface is not covered by the tray?

(StU-14-B3)

TIPS

Tip for Problem 5.1:
Each side of the triangle and the square is a line segment. What is the largest number of points a line segment can have in common with a circle or a square?

Tip for Problem 5.2:
Note that there is no figure offered with this question. The first thing you almost always want to do if you want to figure out a geometric puzzle is to draw a picture of the situation, if you aren't given one anyway. In this particular case, you will see that the answer is really pretty obvious once you've drawn the picture.

Tip for Problem 5.3:
Once again, drawing your own figure is the best path to a solution.

Tip for Problems 5.4, 5.5 and 5.6:
These problems are very similar, and the figure shown for Problem 5.4. applies to all three. It is useful to give names to the radii of the circles, and then to note that (externally) touching circles always have the property that the distance between their centers is equal to the sum of their radii.

Tip for Problem 5.7:
You already know that a figure will be your best tool to get started. Once you have a nice picture, you might want to check it for right angles.

Tip for Problem 5.8:
If a point P lies outside a circle, and the tangents are drawn from P to the circle, the distances from P to the points in which the tangents touch the circle are always of equal length. This is because both the tangents and the points in which they touch the circle are symmetric with respect to the line joining P and the center of the circle. Draw your own figure and take a look! You will find that this is a very useful property to solve this problem.

Tip for Problem 5.9:
There are some interesting right angles in this picture that you might not notice right away. If you can find them (and give good arguments explaining why they must be right angles in the first place), you will be well on your way.

Tip for Problem 5.10:
Try adding the line segment MC to your figure. How long is this segment?

Tip for Problem 5.11:
A tangent of a circle is perpendicular to the radius in the point of tangency. That's really all you need to solve this one.

Tip for Problem 5.12:
There are a few more angles in this figure equal to $\angle MDC$ and $\angle CMP$. If you can find them, you will be well on the way to a solution.

Tip for Problem 5.13:
You might have noticed that we know something about the line segment CC', even though this segment hasn't been drawn in the figure. Try adding it to your figure. What do you notice about the resulting triangle?

Tip for Problem 5.14:
It will help a lot to add two vertical chords of the circles to the figure on the left and on the right. You might then notice some sections of the figure with identical shapes and areas.

Tip for Problem 5.15:
This is another one of those problems, where the lack of a figure in the problem setting should tip you off right away. Draw your own figure and take a close look. There are a lot of similar triangles here, and it will help to identify them.

Tip for Problem 5.16:
Once again, a nice figure will get you half-way there. You may also notice a useful square in there somewhere.

SOLUTIONS

Solution for Problem 5.1:
The largest number of points a line segment can have in common with a circle is two. Since a triangle is made up of three line segments and a square is made up of four, there are seven line segments in play here as sides of these two shapes, and the largest number of points that could conceivably lie both on one of these segments and the circle is equal to $7 \times 2 = 14$. Similarly, the largest number of points a line segment can have in common with a square is also equal to two, and the largest number of points that could conceivably lie both on one of the three sides of the triangle and on the square is equal to $3 \times 2 = 6$. The largest possible number of points of intersection that could result by placing the shapes in an appropriate manner is therefore $14 + 6 = 20$.

We can see that this number is actually attainable by looking at the figure.

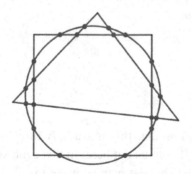

Solution for Problem 5.2:

The figure really says it all. The smallest circle containing both of the tangent circles must have a diameter $2r$ equal to the sum of the diameters of the two smaller circles it contains. This means $2r = 2 \times 3 + 2 \times 5 = 16$, and therefore $r = 8$.

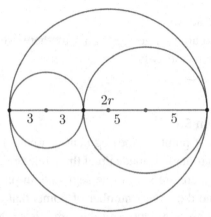

Now, if you are reading this, and the whole idea of a figure "saying it all" about a mathematical argument makes you uneasy, you are, of course, quite right. We are making a silent assumption here that the smallest encompassing circle results from placing the midpoint on the line segment connecting the centers of the two given circles. While this does, in fact, yield the smallest encompassing circle, we really need a proof for this to be a complete mathematical argument. If you know how to construct such an argument, you might want to skip this next part, but if you don't, you may find it quite informative to read on.

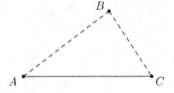

So, how do we show that this is true? A little something called the triangle inequality will help us out here. In this context, this is actually the same as saying that the shortest path between two points is a straight line. In any triangle ABC, the sum of the lengths of sides AB and BC is always greater than the length of the third side AC, since it is a shorter path that leads directly from A to C than the one that makes a detour through B, as we can see in the figure above. The dashed connection $AB + BC$ is certainly longer than the direct solid connection AC.

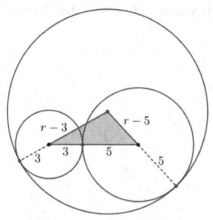

Now, what would happen, if we consider an encompassing circle in our problem, whose center does not lie on the line joining the centers of the two original touching circles? This situation is shown below.

Whenever two circles touch each other, their centers lie on a common line with their common point. As we see in the figure, the centers of the three circles form a triangle with sides of length $3 + 5 = 8$, $r - 3$ and $r - 5$ (letting r denote the radius of the encompassing circle). From the triangle inequality, we see that

$$(r - 3) + (r - 5) > 8$$

holds, which is equivalent to $r > 8$. We see that the smallest possible radius length, namely 8, really does result if we assume the center of the encompassing circle to lie on the line joining the centers of the touching circles, as any other position would result in a radius larger than 8.

Solution for Problem 5.3:

As is so often the case, a good drawing will help us a lot. Noting that the points whose distance from P is equal to 7 cm are precisely the points of the circle with center P and radius 7 cm, we get the figure shown here. Right away, we see that Q can lie either outside k or on k. Of course, just as was the case in the last problem, anyone with an inquisitive and critical way of thinking (which is a really useful thing to have if you want to do

mathematics) will want to know why. The point labeled A in the figure is
the common point of the line PM and k, lying opposite P with respect to
M. The distance from P to this point A is equal to $PM + MA = 2 + 5 = 7$.
So all we need to show now, is that any other point Q with $PQ = 7$ lies
outside of k.

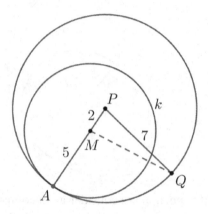

Once again, we can derive this from the triangle inequality. Let us
consider the triangle PMQ. From the triangle inequality, we know that
$PM + MQ > PQ$. Since $PM = 2$ and $PQ = 7$, this is equivalent to
$2 + MQ > 7$, or $MQ > 5$, and this means that Q lies outside of the
circle k, whose radius is 5.

Solution for Problem 5.4:

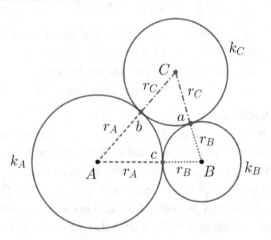

In order to solve problems 5.4, 5.5 and 5.6, it is useful to add some notation to the figure. As we see, circle k_A has the center A and the radius r_A, and we apply analogous labels for the other two circles with centers B and C. In the triangle ABC, the sides each connect the centers of touching circles, and the lengths of the sides are therefore each equal to the sums of two of the radii. We therefore have

$$a = r_B + r_C, \quad b = r_C + r_A \quad \text{and} \quad c = r_A + r_B.$$

Specifically, for Problem 5.4, we are given

$$a + b + c = 26\,\text{cm} \quad \text{and} \quad r_A = 5\,\text{cm},$$

and wish to determine $BC = a = r_B + r_C$. Having defined this notation, it is now quite straightforward to calculate

$$26 = a + b + c = (r_B + r_C) + (r_C + r_A) + (r_A + r_B) = 2(r_B + r_C) + 10,$$

which yields $BC = a = r_B + r_C = (26 - 10) \div 2 = 8\,\text{cm}$.

Solution for Problem 5.5:
Using the figure and the notation introduced in the last solution, we are now given

$$a + b + c = 34\,\text{cm}, \quad r_A = 5\,\text{cm} \quad \text{and} \quad r_B = 8\,\text{cm},$$

and wish to determine r_C. This time, we calculate

$$34 = a + b + c = (r_B + r_C) + (r_C + r_A) + (r_A + r_B) = 2 \times r_C + 2 \times 5 + 2 \times 8,$$

which yields $r_C = (34 - 26) \div 2 = 4\,\text{cm}$.

Solution for Problem 5.6:
This problem is slightly trickier that the last two, even though it also refers to the same situation. Here, we are given

$$a = r_B + r_C = 12\,\text{cm}, \quad b = r_C + r_A = 15\,\text{cm} \quad \text{and} \quad c = r_A + r_B = 13\,\text{cm},$$

and wish to calculate r_C. Simple substitution in the style of the last two solutions will not be quite enough, but we can add all three given equations, which gives us

$$(r_B + r_C) + (r_C + r_A) + (r_A + r_B) = 2(r_A + r_B + r_C)$$
$$= 12 + 15 + 13 = 40\,\text{cm},$$

and therefore $r_A + r_B + r_C = 40 \div 2 = 20\,\text{cm}$. This then yields

$$r_C = 20 - (r_A + r_B) = 20 - 13 = 7\,\text{cm},$$

and we are done.

Solution for Problem 5.7:
As we can see in the figure below, we let M denote the center of the small interior circle, r its radius, P the point in which it touches $M_1 M_2$, and A the point in which it touches k_1.

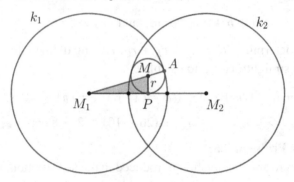

If we take a closer look at the triangle MPM_1, which is shaded in the figure, we can first note that this triangle must be right, as the radius MP of the interior circle must be perpendicular to the tangent $M_1 M_2$. Since A, M and M_1 lie on the same line, the length of the hypotenuse MM_1 of this right triangle is equal to $M_1 A - MA = R - r = 32 - r$. The length of the leg MP is, of course, equal to r, and the length $M_1 P$ of the other leg is half the length of $M_1 M_2$. Since we are given that the points in which the circles intersect $M_1 M_2$ divide it into three parts of equal length, the length of $M_1 M_2$ is equal to $\frac{3}{2} \times R = \frac{3}{2} \times 32 = 48$, and since the length of $M_1 P$ is half of this, we have $M_1 P = 24$. The Pythagorean Theorem in MPM_1 now gives us

$$M_1 M^2 = MP^2 + M_1 P^2, \quad \text{or} \quad (32 - r)^2 = r^2 + 24^2.$$

This can be written in the form $1024 - 64r + r^2 = r^2 + 576$, or $r = (1024 - 576) \div 64 = 7$, and we are done.

Solution for Problem 5.8:

As was already mentioned in the tips section, if the tangents are drawn from some point outside a circle, the distances from this point to the points in which the tangents touch the circle are always of equal length. In order to apply this fact to the given situation, we name some more of the points in the figure, as shown.

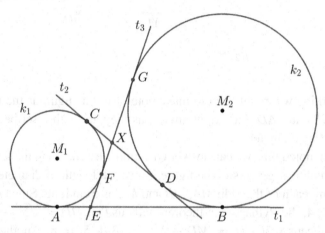

Let F denote the common point of t_3 and k_1, G the common point of t_3 and k_2, and X the point of intersection of t_2 and t_3. Because $AB = AE + EB$, we use the tangent property to obtain $EB = EG$, $AE = EF$, $XC = XF$ and $XG = XD$, and from this we can calculate

$$AB = AE + EB$$

$$= AE + EG$$

$$= AE + (EF + FG)$$

$$= AE + EF + (FX + XG)$$

$$= AE + AE + (CX + XD)$$

$$= 2 \times AE + CD$$

$$= 2 \times 3 + 7 = 13 \, \text{cm}.$$

Solution for Problem 5.9:

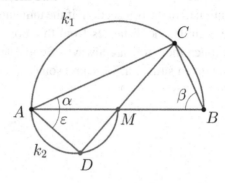

The first thing we might want to make note of in this figure is the fact that both $\angle ACB$ and $\angle ADM$ are right angles, as they are both inscribed angles subtended by diameters.

Having noted this, we can answer Question 1 by checking the lengths of the various line segments in the figure. Since the length of diameter AB of k_1 is 10 cm, each of the radii MA, MB and MC is 5 cm long. Since $MA = 5$ and $AD = 4$, the Pythagorean theorem tells us that $MD = \sqrt{5^2 - 4^2} = 3$, and this means that $CD = MD + MC = 3 + 5 = 8$. Another application of the Pythagorean theorem, this time in the right triangle ACD, gives us

$$AC = \sqrt{AD^2 + CD^2} = \sqrt{4^2 + 8^2} = \sqrt{80}.$$

A final application of the Pythagorean theorem, this time in the right triangle ACB, then gives us

$$BC = \sqrt{AB^2 - AC^2} = \sqrt{10^2 - 80} = \sqrt{20}.$$

This completes the solution to (a).

Continuing on to part (b), we note that the style of this question is now quite different from the last few problems we have considered. Since we are now talking about angles, the right angles we used to solve part (a) can still be relevant, but we will also need to make some observations of a quite different nature. The most important tools will be the fact that sum of the interior angles of a triangle is always equal to $180°$ and the fact that isosceles triangles always have two equal interior angles opposite their equal sides.

In (b), we are first asked to calculate β if we are given $\varepsilon = 45°$. If indeed $\varepsilon = 45°$, we can calculate

$$\angle AMD = 180° - \angle MAD - \angle ADM = 180° - 45° - 90° = 45°,$$

and since vertically opposite angles are certainly equal, we also have $\angle CMB = \angle AMD = 45°$. Noting now that triangle MCD is isosceles, since MC and MD are both radii of k_1, we can use the angles in this triangle to obtain

$$180° = \angle CMB + \angle MCB + \angle MBC = 45° + \beta + \beta,$$

which gives us $\beta = \frac{1}{2}(180° - 45°) = 67\frac{1}{2}°$.

Finally, if we want to find conditions for $\alpha = \varepsilon$ to hold, we can consider the angles in M. As we just saw, we can use the sum of angles in triangle AMD to obtain

$$\angle AMD = 180° - \angle MAD - \angle ADM = 180° - \varepsilon - 90° = 90° - \varepsilon.$$

Noting that triangle MAC is isosceles, since MA and MC are both radii of k_1, we similarly obtain

$$\angle AMC = 180° - \angle MAC - \angle ACM = 180° - \alpha - \alpha = 180° - 2\alpha.$$

Since C, M and D lie on a common line, we have $\angle AMD + \angle AMC = 180°$, and this gives us

$$(90° - \varepsilon) + (180° - 2\alpha) = 180°.$$

If $\alpha = \varepsilon$ holds, this is equivalent to

$$(90° - \varepsilon) + (180° - 2\varepsilon) = 180°, \quad \text{or} \quad \varepsilon = 30°,$$

and we are done.

Solution for Problem 5.10:

Adding the line segment *MC* to the figure, we note that the lengths of *MB*, *MC* and *CP* are all equal to the radius *r* of the circle. This means that triangles *CPM* and *MCB* are both isosceles, and this fact will get us to the solution.

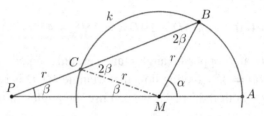

Since the triangle *CPM* is isosceles with $CP = CM = r$, we have $\angle CMP = \angle CPM = \beta$ and $\angle MCP = 180° - 2\beta$. This implies $\angle MCB = 180° - \angle MCP = 2\beta$, and since the triangle *MCB* is also isosceles with $MC = MB = r$, we have $\angle MBC = \angle MCB = 2\beta$ and $\angle CMB = 180° - 4\beta$. Now, all that is left to do is to note that

$$\angle CMP + \angle CMB + \angle AMB = 180°$$

holds, since this gives us

$$\beta + (180° - 4\beta) + \alpha = 180°,$$

which is equivalent to $\alpha = 3\beta$, and since we are given $\alpha = 70°$, we therefore have $\beta = \frac{70°}{3} = 23\frac{1}{3}°$.

Solution for Problem 5.11:

Since *AB* and *AD* are tangents of *k*, we have $\angle ABC = \angle ADC = 90°$.

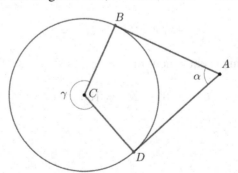

Since the sum of the angles in the quadrilateral $ABCD$ is equal to $360°$, we therefore obtain

$$\angle BCD = 360° - \angle ABC - \angle ADC - \angle BAD$$

$$= 360° - 90° - 90° - \alpha = 180° - \alpha,$$

and since $\angle BCD + \gamma = 360°$, and γ is five times as big as α, we therefore obtain

$$(180° - \alpha) + 5\alpha = 360°,$$

which is equivalent to $4\alpha = 180°$ or $\alpha = 45°$.

Solution for Problem 5.12:
We are given that the angles $\angle MDC$ and $\angle CMP$ are equal, so let's call this angle α. Since C and D both lie on the circle with center M, we have $MD = MC$, and since the triangle MCD is therefore isosceles, we also have $\angle MCD = \angle MDC = \alpha$. Also, since CP is parallel to MD, we also have $\angle PCD = \angle MDC = \alpha$. This, of course, gives us $\angle MCP = 2\alpha$.

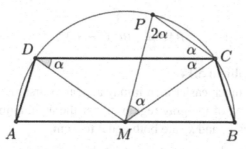

Now, all that is left is for us to note that $MP = MC$ must also hold, since P also lies on the circle. The triangle MPC is therefore also isosceles with $\angle MPC = \angle MCP = 2\alpha$. Since the sum of the interior angles of this triangle is equal to $180°$, we obtain

$$\angle PMC + \angle MCP + \angle MPC = 180,$$

and this is equivalent to

$$\alpha + 2\alpha + 2\alpha = 5\alpha = 180°,$$

or $\alpha = 36°$.

Solution for Problem 5.13:

On the one hand, we are given that C and C' are symmetric with respect to AB, and this gives us $BC = BC'$. On the other hand, we are also given that $BC = CC'$ holds. The triangle BCC' is therefore equilateral, and all its interior angles are equal to 60°.

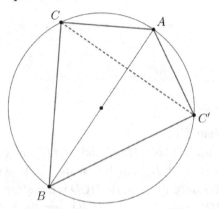

Since the angles subtended by the arc BC at A and C' are equal, we therefore have

$$\angle BAC = \angle BC'C = 60°.$$

Solution for Problem 5.14:

This problem is much easier than it may appear at first glance. We wish to determine the area of the gray region, under the assumption that the radii of the two circles k_1 and k_2 are both equal to 2 cm.

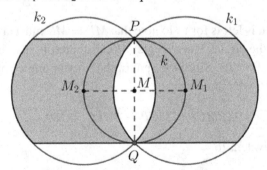

We let A, B, C and D denote the points in which the horizontal lines through P and Q intersect with k_1 and k_2, as shown below. As shown in the figure on the right, the shaded regions are identical, since a horizontal translation transforms one onto the other. Since the same also holds on the right-hand side of the figure, the segments that AB and CD cut off the circles k_1 and k_2 have the same area as the inside non-shaded portion of the figure we are considering. In other words, the area of the gray region is equal to the area of the rectangle $ABCD$.

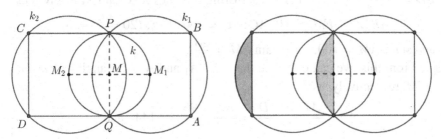

All we need to do is therefore to calculate the area of this rectangle, but that isn't too difficult, either. Since PQ and $M_1 M_2$ are perpendicular diameters of k, PM_1QM_2 is a square and since M_1 is the center of k_1 and M_2 is the center of k_2, $ABPQ$ and $CDQP$ must also be squares. The lengths of the sides of these squares are equal to $\sqrt{2}$ times the radii of the circumscribed circles, i.e. $2\sqrt{2}$ cm. The area of each of these squares is therefore equal to $(2\sqrt{2})^2 = 8\,\text{cm}^2$, and the area of the rectangle $ABCD$, which is equal to the area of the gray region, is therefore twice as large, and therefore equal to $16\,\text{cm}^2$.

Solution for Problem 5.15:
Adding the line segment BC to the figure, we note that the triangle ABC is right, since the angle subtended by the diameter AB in C is certainly a right angle. By the Pythagorean theorem, we therefore have $CB = \sqrt{200^2 - 160^2} = 120$.

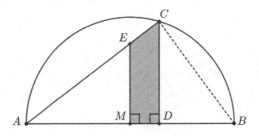

Now, we can readily see that the right triangles ABC, ACD and AEM are certainly similar as they all have a common angle in A. Comparing triangles ABC and ACD therefore gives us

$$AC : AD : CD = AB : AC : BC = 200 : 160 : 120 = 5 : 4 : 3,$$

and since $AC = 160$, we therefore obtain $AD = \frac{160 \times 4}{5} = 128$ and $CD = \frac{160 \times 3}{5} = 96$. Similarly, comparing triangles ABC and AEM gives us

$$AE : AM : EM = AB : AC : BC = 200 : 160 : 120 = 5 : 4 : 3,$$

and since $AM = 100$, we obtain $EM = \frac{100 \times 3}{4} = 75$.

From this, we can calculate the areas A_1 and A_2 of triangles ACD and AEM, respectively, as

$$A_1 = \frac{AD \times CD}{2} = \frac{128 \times 96}{2} = 6144 \quad \text{and}$$

$$A_2 = \frac{AM \times EM}{2} = \frac{100 \times 75}{2} = 3750,$$

which yields the required area of $MDCE$ as $A_1 - A_2 = 6144 - 3750 = 2394$.

Solution for Problem 5.16:

Taking a look at the figure, we see that A, M and B are vertices of a square inscribed in the circle of the tray, as MA and MB are equally long and perpendicular. If we assume that the radius of the table is equal to 1 and let N denote the center of the tray and A_1 the area of each of the circular segments on the tray over AM and BM, we see that the area of the table's whole surface is equal to $1^2 \times \pi = \pi$, and the part not covered by the tray is equal to $\frac{3}{4}\pi - 2A_1$.

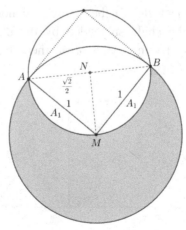

All that remains is for us to calculate the value of A_1. Since N is the center of both the circular tray and the inscribed square, the sides of the square must be $\sqrt{2}$ times as long as the radius $NA = NM$. We therefore have $NA = \frac{1}{\sqrt{2}} = \frac{\sqrt{2}}{2}$, and A_1 is therefore equal to one quarter the area of the circular tray minus the area of the right isosceles triangle ANM. In other words, we have

$$A_1 = \frac{1}{4} \times \left(\left(\frac{\sqrt{2}}{2} \right)^2 \times \pi \right) - \frac{1}{2} \times \left(\frac{\sqrt{2}}{2} \right)^2 = \frac{1}{8} \times \pi - \frac{1}{4}.$$

We can therefore calculate the area of the exposed part of the table as

$$\frac{3}{4}\pi - 2 \times A_1 = \frac{3}{4}\pi - \frac{1}{4}\pi + \frac{1}{2} = \frac{1}{2}(\pi + 1),$$

and the required fraction is therefore equal to

$$\frac{\pi + 1}{2\pi}.$$

Chapter 6

Logic

Some problems don't really require any mathematical knowledge at all to solve. (That is, unless you include counting as mathematical knowledge.) All it takes to solve one of these gems is some insight into the situation and some argumentative savvy. One type of such problems deals with people who are known to tell the truth or known to lie. Another type asks us to find the number of objects with some specific obscure trait, while others deal with equally unlikely situations. What they all have in common is the fact that pure logical reasoning will get you to a solution, without requiring any serious calculation.

PROBLEMS

Problem 6.1:
Albert, Bruno, Christoph and Dieter are under suspicion of having broken a window. When they are questioned about this, Albert and Bruno each state "It wasn't me!". Dieter says "It was Bruno!". Christoph says "It was Dieter!". It is known that exactly one of the four is lying. Who broke the window?

(StU-01-B3)

Problem 6.2:
There are two villages in the Schmähführ valley, Blunz and Brein. The inhabitants of both villages tell the absolute truth on four days of the week, but the Blunzers always lie from Monday through Wednesday and the Breiners always lie from Thursday through Saturday. Sepp and Hans, one of whom is from Blunz and one of whom is from Brein, meet up for a sausage

lunch one day. Sepp says "I lie on Saturday". Hans answers "I am going to lie tomorrow". Finally, Sepp says "I lie on Sundays, too". Which of the two is from Blunz? On which day of the week are they having their sausages?

(StU-04-B4)

Problem 6.3:

Of the following six statements about the four participants in a contest, exactly one is false, while all the others are true:

(i) Alex scored more points than Conny.
(ii) Betty scored less points than Dan.
(iii) Conny scored more points than Dan.
(iv) Alex scored less points than Betty.
(v) Betty and Conny together scored twice as many points as Alex.
(vi) Conny scored the mean of the numbers of points scored by Alex and Dan.

Which of the six statements is false? Arrange the participants according to their scores.

(StU-12-B3)

Problem 6.4:

There are hooks on the wall near the door of a restaurant for people to hang their coats. The hooks are attached at regular intervals. Stinky Henry arrives at the restaurant and hangs his foul-smelling overcoat on one of the hooks. Any other guests must leave at least three empty hooks between their coats and his to avoid attracting the stench. Soon afterward, the twin sisters Linda and Belinda arrive, each wearing a thick fur coat. These are so thick that no other guest can hang his or her coat right next to it. It turns out that only six more patrons will be able to hang their coats after this. What is the largest possible number of hooks on the wall?

(StU-13-A10)

Problem 6.5:

On the planet Xxu, there are Aazes, Bbehs und Zeezes. Each resident of the planet has at least one head. If someone has only one head, he is called an

Ainzie, and is pitied by the others because of his loneliness. All Aazes have the same number of heads. All Bbehs and Zeezes have less heads than the Aazes, but not necessarily the same number. There are 5 Aazes, 3 Bbehs and 4 Zeezes standing in a group, singing a song in 8-part harmony. There is no Ainzie among them, and together they have 36 heads. How many heads does each Aaz have?

(StU-14-B2)

Problem 6.6:
If Anna is borr, she is certainly gemm. If Anna is not fatt, she is certainly not zokk. Anna is certainly either zokk or not gemm (but possibly both!). After careful consideration, it turns out that Anna is borr. Is she also fatt?

(StU-14-B4)

Problem 6.7:
Astrid goes to the county fair and buys a packet of chocolate sticks. The packet contains four different types of sticks: five are milk chocolate and five are coconut. Four are rum chocolate and four are rum coconut. She wants to eat a stick containing coconut first, but she can't tell what kind she has in her hand until she takes it out of the packet. How many does she have to take out of the packet so that she is sure to have one with coconut among them?

(O-02-5)

Problem 6.8:
Ms Schlamperl has a box in her attic containing various woolen mittens that cannot be differentiated by touch. 3 are left red mittens, 6 are left green mittens, 4 are right red mittens and 6 are right green mittens. There is no light in the attic, and she has to take the mittens out of the box blindly. What is the smallest number of mittens she has to take out of the box in order to be sure of having a left and a right mitten of the same color?

(StU-98-A4)

Problem 6.9:
There are 7 blue balls, 9 red balls and 10 white balls in a container. Balls are taken out of the container without looking at them, until either three balls

of the same color have been taken out, or one ball of each color. What is the largest number of balls that must be taken out for this to be the case?

(O-94-7)

Problem 6.10:
Juanita is a student at a school with an intense modern language program. There are 40 students in her class, and each of them takes at least one of the modern languages English, German or French: 34 of them take at least one of the languages English or German, 25 of them take at least one of the languages German or French. 6 students only take German. The combination of only English and German is taken by three students more than the combination of only French and German. No students take the combination of English and French. How many students take exactly one language? How many take exactly two?

(StU-17-B6)

Problem 6.11:
Magic Michaela has four arms. She has a sweater with seven sleeves. When she puts the sweater on, she can use any combination of four sleeves for her arms, but always exactly four. Whenever she has used a sleeve ten times, it falls off. What is the largest number of times she can wear the sweater, assuming she can never show a bare arm in public?

(StU-16-B6)

Problem 6.12:
On a hike, mother, father, son and daughter come to a river which they need to cross. There is a rowboat tied up at the crossing, which can transport a maximum weight of 75 kg at a time. The mother weighs 70 kg, the father weighs 75 kg and each of the siblings weighs 35 kg. Each of the family members is equally capable of rowing the boat to the other side. What is the smallest number of river crossings with which the family can get to the other side?

(O-92-4)

TIPS

Tip for Problem 6.1:
What do they each have to say about Bruno's role in all of this?

Tip for Problem 6.2:
Who lies on Sundays?

Tip for Problem 6.3:
Focus on the first four claims. Try to find statements that are certainly correct to start with.

Tip for Problem 6.4:
How many hooks are useless due to the strange properties of the hanging coats?

Tip for Problem 6.5:
How many heads must each of the Bbehs und Zeezes have if the Aazes each have 2 heads? What if they each have 3, or 4, or 5?

Tip for Problem 6.6:
Huh? Ok, so this one is a bit strange. Obviously, each of the nonsense words is meant as an adjective, describing some property Anna may or may not have. We are told outright that she has one of these weird properties, and starting from there should get you to an answer.

Tip for Problem 6.7:
What is the largest number of sticks she could take out without getting a coconut stick?

Tip for Problem 6.8:
Think backwards. What is the largest number of mittens she can get without having a useful pair?

Tip for Problem 6.9:
This one is not too far away from the last two. Once again, it will be useful to think how many balls might be drawn without getting a set of one of the required types.

Tip for Problem 6.10:
You might consider making a drawing here. Just draw a circle containing all the students taking English, another containing all the students taking German, and a third containing all the students taking French.

Tip for Problem 6.11:
There are two parts to this. First, you might want to consider what it takes for an arm to fall off. How can she shuffle the sleeves to make the most of her sweater? Once you come up with a number, you will need to find out whether this number of times is actually attainable.

Tip for Problem 6.12:
There is nothing to be gained by one of the family members rowing back and forth on their own, but there is only pair of family members that can cross the river together.

SOLUTIONS

Solution for Problem 6.1:
Note that Dieter claims that Bruno broke the window, while Bruno claims that he did not. Obviously, one of these two is lying. Since we know that only one of the four is lying, the other two are telling the truth. From this, we know that Albert did not break the window, as he says that he didn't, but we also know that Christoph told the truth when he said that Dieter was the culprit who broke the window. Incidentally, we also know that Dieter was the liar, since he tried to pin it on Bruno, which is not really surprising considering his own role in the whole mess.

Solution for Problem 6.2:
We know that Blunzers lie from Monday through Wednesday and that Breiners lie from Thursday through Saturday. Since they all tell the truth on other days, none of them lie on Sunday, and this means that Sepp is lying when he claims to do just this. From this, we know that Sepp is lying today, and he is therefore also lying when he claims to lie on Saturday. If he did lie on Saturday, he would be a Breiner, but since he doesn't, he must be a Blunzer. From all this, we know both that Hans is a Breiner, and that he is telling the truth today, since there is no day on which both Blunzers and Breiners lie. He is therefore telling the truth when he says that he will be lying tomorrow, and since he is telling the truth today, it must be Wednesday, since that is the only day on which a Breiner tells the truth that is followed by a day on which he lies.

Solution for Problem 6.3:

Some notation will help us here. We write A for the number of points Alex scored, and B, C and D for the point scores of Betty, Conny and Dan, respectively. It will be a bit easier if we write the six statements mathematically:

(i) $A > C$ (ii) $D > B$ (iii) $C > D$ (iv) $B > A$
(v) $2A = B + C$ (vi) $2C = A + D$.

If the first four statements were all true, we would have $A > C > D > B > A$, which is obviously not possible. We therefore know that one of these four statements is false, which means that the other two must be true. We know from (v) that A lies between B and C and we know from (vi) that C lies between A and D. This means that the order of the scores must be $B > A > C > D$, and the false statement must be (ii).

Solution for Problem 6.4:

If Stinky Henry's overcoat is hanging somewhere in the middle, there are 3 unusable hooks on either side of it, while there would be less if it were hanging on the side. In effect, his stench can be blocking a total of at most 7 hooks, namely the one on which his overcoat hangs and three on either side. Similarly, each of the twins' coats make a total of at most 3 hooks unusable, namely the one it hangs on and one on either side. Therefore, the three coats already hanging are blocking a total of at most $7 + 3 + 3 = 13$ hooks. Since there are still six usable hooks left, there are at most $13 + 6 = 19$ hooks on the wall.

Solution for Problem 6.5:

Since the Bbehs and Zeezes each have at least 2 heads, and the Aazes have more, they must have at least 3 heads each. If they had 3 heads each, the 5 Aazes would have a total of $3 \times 5 = 15$ heads, leaving $36 - 15 = 21$ heads for the $3 + 4 = 7$ Bbehs and Zeezes. In this case, at least one of them must have at least three heads (more, in fact), which contradicts the fact that the Aazes each have more heads. It therefore follows that the Aazes each have at least 4 heads.

Now, let us assume that the Aazes each have 5 (or more) heads. In this case, the 5 of them have at least $5 \times 5 = 25$ heads altogether, leaving at most only $36 - 25 = 11$ heads for the 7 Bbehs and Zeezes. This would contradict the fact that there are no Ainzies amongst them, however, as they must have at least a total of $7 \times 2 = 14$ heads.

We therefore see that each Aaz must have 4 heads.

Solution for Problem 6.6:
We are told outright in the last statement that Anna is in fact, borr. In the first statement, we are told that this means that she is also gemm. Since she is either zokk or not gemm, but we have already established that she is, in fact, gemm, she must also be zokk. It therefore follows that she must be fatt, since we know that she would not be zokk if she were not fatt. Phew.

Solution for Problem 6.7:
This is really pretty obvious when you think about it a bit. There are 5 milk chocolate sticks and 4 rum chocolate, for a total of $5 + 4 = 9$ non-coconut sticks. All the others contain coconut. If she takes nine or less, she may not have a coconut stick among them, but if she takes 10, at least one of them must contain coconut.

Solution for Problem 6.8:
If Ms Schlamperl is very unlucky, she could grab all of the 6 left green mittens and all of the 4 right red mittens. This means that she could take $6 + 4 = 10$ mittens without getting a usable pair. (We call a pair *usable* if it consists of a left and right mitten of the same color.) If she takes 11 mittens out of the box, there are two possibilities. If could have a left green mitten and a right green mitten among them, in which case she would certainly have a usable pair. If not, all 11 of the mittens she takes are either from the three categories left-red, left-green, right-red or left-red, right-red, right-green. In either case, at least one mitten must be from each of the three categories, as the total numbers of mittens in any two of these categories is always smaller than 11. In either case, she would therefore definitely have taken a usable red pair.

The smallest number of mittens she must take in order to be sure of obtaining a usable pair is therefore 11.

Solution for Problem 6.9:
It is possible to take out four balls without three of them being the same color or one ball of each color being among them. Just imagine 2 blue balls and two red balls, for instance. On the other hand, if five balls are taken, and no three of them are of the same color, there can only be two each of one color and the second color, so the fifth must be of the third color. We

see that the largest number of balls we may have to take out of the container is five.

Solution for Problem 6.10:

Since 34 students take English or German, there are $40 - 34 = 6$ that take only French. Similarly, since there are 25 that take either German or French, there are $40 - 25 = 15$ that take only English. We are given that there are 6 that take only German, and this means that there are $6 + 15 + 6 = 27$ that take only one modern language. On the other hand, we are given that no student takes the combination of English and French, and this means that there is no students that takes all three languages. It therefore follows that $40 - 27 = 13$ students take two languages.

A good way to visualize this is with a diagram of the following type.

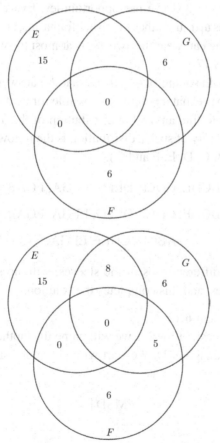

The areas inside the circles labeled E, G and F represent the students taking the languages English, German and French, respectively. The over-lapping areas represent the students taking multiple languages, and the meaning of the numbers should be pretty obvious. Note that we did not use all the information given to solve the problem, but we can fill up the two empty sections by noting that the combination of English and German is taken by 3 students more than the combination of German and French, while the total of these two numbers is already known to be 13. This means that 8 students take English and German while 5 take German and French, and we can complete the diagram as shown below.

Solution for Problem 6.11:

She can use each of the seven sleeves ten times, which gives her a total number of $7 \times 10 = 70$ sleeve-use opportunities. Every time she puts the sweater on, she uses up four of these with her four arms, and so the maximum number of times she can wear the sweater is at most the whole number part of $70 \div 4$, which is 17.

Of course, this reasoning only tells us that the answer cannot be larger than 17. In order to see that it is actually possible for her to wear the sweater 17 times, we need to find a possible distribution of the uses of the various sleeves. One possibility of such a distribution is the following (we name the seven sleeves A, B, C, D, E, F and G):

ABCD, BCDE, CDEF, DEFG, EFGA, FGAB, GABC,

ABCD, BCDE, CDEF, DEFG, EFGA, FGAB, GABC,

ABCD, CDEF, EFGA.

Here we have 17 different uses of the sleeves, with no sleeve being used more than ten times, and this completes the solution.

Solution for Problem 6.12:

To make this a bit easier to write, we will name the mother M, the father F, the son S and the daughter D. We start with all on one side, and write this situation as

FMSD |

If the first crossing is made by any individual, he or she would have to come right back again, and nothing would be gained. Due to the weight restrictions, the only possible start therefore results in

$$FM \mid SD.$$

One of the siblings, say the son, must then return, resulting in

$$FMS \mid D.$$

Now, there is no way for any pair to cross, and the son crossing alone just brings us back to the situation we had before. The next step is therefore for one of the parents to cross, say the mother, resulting in

$$FS \mid MD.$$

If the mother returns with the boat, we have just reversed the situation, and it is therefore necessary for the daughter to return the boat, resulting in

$$FSD \mid M.$$

If the father now crosses, either the father or mother would have to return again, and once again, nothing would be gained. Similarly, son or daughter crossing alone would only result in a situation we have already had. The only option is for son and daughter to cross together, and one of them to bring the boat back for the father, and so the next situations are

$$F \mid MSD \quad \text{and} \quad FD \mid MS,$$

assuming the daughter brings the boat back. Now the father can cross, yielding

$$D \mid FMS,$$

and the son can go back and cross together with the daughter, yielding

$$SD \mid FM \quad \text{and} \quad \mid FMSD,$$

for a total of nine crossings.

Chapter 7

Fractions

A lot of people are wary of fractions. Sometimes the problem is that they simply find it difficult to understand what they mean. A number on top and a number on the bottom ... what is that exactly? Sometimes it has to do with the calculations involved ... how do you get a handle on them? The following problems all deal with fractions in some way. The first few are pure calculations, but all of them have a twist of some kind, and the answers are always surprisingly simple once you understand the structure. You will be asked to "simplify" or to "calculate". Both of these terms simply mean that you are meant to find a way of expressing the resulting number in as simple a way as possible, which usually involves very small integers. The next group of problems deals with relative sizes of fractions with different numerators and denominators (recall that those are the numbers on top and on the bottom, respectively). Finally, you will find a mixed bag of problems, all of which deal with fractions in some unique way. Some of these problems are posed in a multiple choice format. Just doing the calculations will get you to the answer, of course, but you may try guessing the right answer first. Once you calculate the correct answer, you may be quite surprised!

PROBLEMS

Problem 7.1:

Which of the following fractions is equal to x if we are given

$$\frac{1}{2} + \frac{2}{40} + \frac{3}{600} + \frac{4}{8000} = x?$$

(A) $\frac{1}{2222}$ (B) $\frac{1000}{2222}$ (C) $\frac{1110}{2000}$ (D) $\frac{1111}{2220}$ (E) $\frac{1111}{2000}$

(StU-16-A1)

Problem 7.2:

Simplify the expression

$$\left(1 - \frac{1}{3}\right) \times \left(1 + \frac{1}{4}\right) \times \left(1 - \frac{1}{5}\right) \times \left(1 + \frac{1}{6}\right) \times \left(1 - \frac{1}{7}\right) \times \left(1 + \frac{1}{8}\right) \times \left(1 - \frac{1}{9}\right).$$

(StU-01-A6)

Problem 7.3:

Which of the following fractions is equal to the product

$$\left(1 + \frac{1}{2}\right) \times \left(1 - \frac{1}{2}\right) \times \left(1 + \frac{1}{3}\right) \times \left(1 - \frac{1}{3}\right) \times \left(1 + \frac{1}{4}\right)$$

$$\times \left(1 - \frac{1}{4}\right) \times \left(1 + \frac{1}{5}\right) \times \left(1 - \frac{1}{5}\right) \times \left(1 + \frac{1}{6}\right)?$$

(A) $1 - \frac{1}{4 \times 9 \times 16 \times 25 \times 6}$ (B) $\frac{3}{5}$ (C) $\frac{7}{10}$ (D) $\frac{2}{3}$ (E) $\frac{5}{6}$

(StU-05-A6)

Problem 7.4:

Calculate the value of

$$1 + \frac{1}{9 + \frac{1}{9 + \frac{1}{7}}}.$$

(StU-97-A10)

Problem 7.5:

Calculate the value of

$$\frac{6 + \frac{1}{2}}{1 - \frac{1}{1 + \frac{2}{1 + \frac{1}{1}}}}.$$

(S5K-05-4)

Problem 7.6:

Two-thirds of half of $7 \times A$ is equal to 15. What is the value of three-quarters of a third of $28 \times A$?

(S5K-12-6)

Problem 7.7:

Which fraction with the denominator 17 has a value larger than $\frac{1}{4}$ but smaller than $\frac{1}{3}$?

(StU-91-4)

Problem 7.8:

Which of the numbers 6, 7, 8 or 9 can be substituted for m in the fraction $\frac{m}{50}$ in such a way that the value of the fraction is greater than $\frac{1}{8}$ but less than $\frac{1}{7}$?

(StU-00-A1)

Problem 7.9:

A fraction has the following properties: Its value is greater than $\frac{3}{4}$, but less than $\frac{4}{5}$. Its numerator is a multiple of 6, and its denominator is 5 greater than its numerator. Determine the sum of its numerator and its denominator.

(StU-07-A7)

Problem 7.10:

The same number can be put in place of both squares in the expression

$$\frac{\blacksquare + 3}{12} = \frac{9 + \blacksquare}{20}.$$

Which of the following numbers 4, 5, 6, 7 and 8 can that be?

Are there any other numbers that could replace the squares and yield a true expression?

(StU-13-A4)

Problem 7.11:

Two two-digit numbers are written, using the four digits 2, 3, 4 and 6. The two-digit numbers are written as the numerator and the denominator of a fraction, which is then reduced as far as possible. Which of the following numbers cannot be the result of this process?

(A) $\frac{1}{2}$ (B) $\frac{2}{3}$ (C) $\frac{5}{6}$ (D) $\frac{6}{7}$ (E) $\frac{13}{17}$

(StU-14-A4)

Problem 7.12:

What is the value of $\frac{a^2}{b^2}$ if we are given

$$\frac{a+b}{a-b} = \frac{7}{4}?$$

(StU-96-A3)

Problem 7.13:

Which of the following expressions assumes the smallest value if a random integer greater than 5 is inserted for x? Or does this depend on the concrete values of x we use?

$$\frac{5}{x}, \quad \frac{5}{x+1}, \quad \frac{5}{x-1}, \quad \frac{x}{5}.$$

(StU-96-A9)

Problem 7.14:

How many fractions with the denominator 25 exist, whose value is greater than ¼ but less than ½?

(StU-04-A2)

Problem 7.15:

A unit fraction is a fraction with numerator 1. The number one can be written as the sum of three different unit fractions in the form $1 = \frac{1}{2} + \frac{1}{3} + \frac{1}{6}$. It can also be written as the sum of five different unit fractions: $1 = \frac{1}{2} + \frac{1}{12} + \frac{1}{18} + \frac{1}{A} + \frac{1}{B}$. Determine the smallest possible value of $A + B$.

(S5K-16-3)

Problem 7.16:

Certain sets $\{a, b, c, d\}$ of positive integers have the property that $\frac{1}{a} + \frac{1}{b} = \frac{1}{c} + \frac{1}{d}$ holds. For instance, we have (i) $\frac{1}{12} + \frac{1}{28} = \frac{1}{14} + \frac{1}{21}$ and (ii) $\frac{1}{30} + \frac{1}{95} = \frac{1}{38} + \frac{1}{x}$.

(a) Demonstrate the validity of (i).
(b) Determine the value of x in (ii).
(c) Determine another set of four different positive integers fulfilling this condition.
(d) Prove that an infinite number of such sets exist.

(StU-15-B4)

Problem 7.17:

A traveler fell asleep after half of his trip and woke up again when a distance still remained to his goal that was half as long as the distance he had slept through. What fraction of his entire trip did he sleep through?

(StU-89-5)

Problem 7.18:

Klaus runs twice as fast as he walks. On his way to school, he runs two-thirds of the time and walks one-third of the time, and this takes him a total of 16 minutes. How long will it take him to get home if he walks two-thirds of the time and runs one-third of the time?

(StU-95-12)

Problem 7.19:

At the start of the school year, $\frac{3}{5}$ of the students in a class were girls. After a while, four new students join the class: two girls and two boys. Which of the following claims is certainly true?

  (i) There are equally many boys and girls in the class.
 (ii) There are more boys than girls in the class.
(iii) There are less boys than girls in the class.
 (iv) $\frac{7}{9}$ of the students in the class are girls.
  (v) $\frac{5}{9}$ of the students in the class are girls.

(StU-98-A9)

Problem 7.20:

Snarlotte the snail is climbing up a beanstalk at a constant speed. At 13:00, she has gone $\frac{1}{3}$ of the way up, and at 14:30 she has gone $\frac{7}{12}$ of the way up. What fraction of the total distance up has she gone up at 13:30?

(StU-99-A5)

Problem 7.21:

We have 4.5 liters of a mixture consisting of seven-ninths of water and two-ninths of fruit juice. By adding water, the water content of the mixture is to be increased to eight-ninths. How many liters of the mixture will we have after we have done this?

(StU-07-B6)

118   *Mathematical Nuggets from Austria*

Problem 7.22:
Joe has a 1-liter bottle filled with apple juice. Every day, he drinks half a liter out of the bottle and then refills the empty half of the bottle with water. How much of the apple juice originally in the bottle did Joe consume within the first 4 days?

(StU-09-A1)

Problem 7.23:
Achmed is sitting in front of a big bowl of cherries. He eats 75% of them, and then decides that it is time for a bathroom break. After returning to the bowl, he eats $\frac{5}{6}$ of the remaining cherries, leaving 4 in the bowl. How many cherries were there in the bowl at the beginning?

(StU-14-A1)

Problem 7.24:
A mathematician gives his heirs a final puzzle to solve by writing the following in his will:

"To my sister Claudia, I leave € 600 and a third of the rest of my estate. To my brother Franz I leave € 950 and a quarter of my estate. To my sister Katharina, I leave one-sixth of my estate. Together, this is all I have in the world".

How much money does each of the heirs receive?

(S5K-01-8)

Problem 7.25:
A certain number of questions of equal value are posed on a test. If I answer 9 of the first 10 correctly, and exactly $\frac{3}{10}$ of the remaining ones, I will have exactly 50% of the answers correct. How many questions are posed on the test?

(O-93-5)

TIPS

Tip for Problem 7.1:
We start this chapter with the addition of fractions, which is just the thing that gets some peoples' teeth chattering right off the bat. Recall that you can only add fractions with a common denominator. If you look at these

numbers carefully, you should see a really easy common denominator to work with.

Tip for Problem 7.2:
Just note that you can write 1 as

$$1 = \frac{2}{2} = \frac{3}{3} = \frac{4}{4} = \frac{5}{5} = \cdots$$

and then calculate the value of each of the expressions in parentheses. Don't forget that you can make multiplying fractions much easier if you cancel common factors as soon as they pop up.

Tip for Problem 7.3:
Remember what we said about the last two problems. As already mentioned in the first problem, it is probably a good idea to ignore the suggested answers and just do the calculations yourself. You will see that you get one of these suggested answers anyway. Also, the structure of the given expression is an awful lot like the one in the second problem, so you may want to start off with the same type of steps.

Tip for Problem 7.4:
It is pretty obvious, which year this problem was first posed in. If you want to simplify a continued fraction, it is often a good idea to expand the big fraction by the denominator of the small one. For example, you can write

$$\frac{1}{1 + \frac{1}{2}} = \frac{2 \times 1}{2 \times \left(1 + \frac{1}{2}\right)} = \frac{2}{2 + 1} = \frac{2}{3}.$$

Tip for Problem 7.5:
This one barely qualifies as a continued fraction. Just look at from the bottom up, and you will see what is going on here.

Tip for Problem 7.6:
This may just be a bit confusing because it is written out. Try "translating" both the first and the second part to algebraic expressions; it will make things much easier.

Tip for Problem 7.7:
Can you write the numbers $\frac{1}{4}$ and $\frac{1}{3}$ as seventeenths? Try finding a common denominator first.

Tip for Problem 7.8:
Take a look at the solution of the last problem. The same method will work here.

Tip for Problem 7.9:
Once again, an inequality will help. Try defining a multiple of 6 as $6x$.

Tip for Problem 7.10:
The easiest way to find out, which of the given numbers can be put in place of the squares is to try each of them out in turn. That won't answer the second part of the question, though. But you can deal with that part by treating the squares as variables; try writing x instead of a square.

Tip for Problem 7.11:
There are only six possible combinations of two digits you can obtain to write two-digit numbers in this way. That means a total of 12 such two-digit numbers. You can write down all these pairs, cancel the resulting fractions, and see what happens.

Tip for Problem 7.12:
Try multiplying the equation with the product of the denominators. What does this tell you about the ratio $a{:}b$?

Tip for Problem 7.13:
You will have to decide whether this depends on the concrete values of x or not, but you can try out a few numbers to get a feeling. Try out what happens for $x = 6$. Then try $x = 7$ or $x = 8$. This should give you an idea. Can you prove it?

Tip for Problem 7.14:
If you let x stand for the numerator of such a fraction, you should be able to set up inequalities. As we have seen in a few problems already, solving inequalities involving fractions can sometimes be dealt with quite efficiently by finding a common denominator.

Tip for Problem 7.15:
Try treating this as a regular equation in two variables. Eliminate the fractions by multiplying with something appropriate, and then don't forget that A and B are positive integers!

Tip for Problem 7.16:
The first two parts are simple calculations, but they are meant to give you a starting point for parts (c) and (d). Think of what you can do with an equation known to be true that changes the numbers but keeps the equation true.

Tip for Problem 7.17:
Think of the length of the piece remaining after he wakes up as x. How long was the piece he slept through? How long was the piece before that?

Tip for Problem 7.18:
If Klaus walks x meters in t minutes and runs twice as fast as he walks, he runs $2x$ meters in t minutes.

Tip for Problem 7.19:
You can figure this one out by trying different values for the number of boys and girls in the class. What if there were 3 girls at the beginning? What if there were 6?

Tip for Problem 7.20:
What fraction of the total distance has she gone up between 13:00 and 14:30? What part of that has she gone up between 13:00 and 13:30?

Tip for Problem 7.21:
Try adding x liters of water. The resulting equation should do the trick.

Tip for Problem 7.22:
Note that he is drinking half of the bottle's contents every day. That means he is also drinking half of the remailing apple juice every day.

Tip for Problem 7.23:
This is another problem that can be easily attacked with a standard equation. You want to know how many cherries there were in the bowl, so just call that number x and write the statement as an equation.

Tip for Problem 7.24:
This is quite similar to the last problem, and a standard type of equation will do quite nicely here as well. Try letting x denote the total value of the estate.

Tip for Problem 7.25:
This last problem of the section is another one you can solve with an equation. This time, just let x denote the unknown number of questions on the test.

SOLUTIONS

Solution for Problem 7.1:
Take a close look at the four fractions. There are some really obvious common factors you can cancel, as the numerators are all divisors of the respective denominators. This means that we immediately get

$$\frac{1}{2} + \frac{2}{40} + \frac{3}{600} + \frac{4}{8000} = \frac{1}{2} + \frac{1}{20} + \frac{1}{200} + \frac{1}{2000}.$$

It should now not escape your attention that all of the denominators start with the digit 2. The largest of them, namely 2000, is therefore the smallest common denominator, as we simply multiply each of the others by some power of 10 to obtain it. Since we can do the same with the numerators by just adding on the same number of zeros missing from the denominators to get 200, we have

$$\frac{1}{2} + \frac{1}{20} + \frac{1}{200} + \frac{1}{2000} = \frac{1000}{2000} + \frac{100}{2000} + \frac{10}{2000} + \frac{1}{2000},$$

and adding the numerators gives us

$$\frac{1000}{2000} + \frac{100}{2000} + \frac{10}{2000} + \frac{1}{2000} = \frac{1111}{2000}.$$

Note that this problem was originally used at a multiple-choice math competition. The alternative solutions offered may seem plausible at first glance, but ignoring them and simply doing the calculation in your head is a much better tactic than worrying about them at all, at least in this case.

Solution for Problem 7.2:
Since we know that

$$1 = \frac{2}{2} = \frac{3}{3} = \frac{4}{4} = \frac{5}{5} = \cdots$$

holds, we can write

$$\left(1-\frac{1}{3}\right) \times \left(1+\frac{1}{4}\right) \times \left(1-\frac{1}{5}\right) \times \left(1+\frac{1}{6}\right) \times \left(1-\frac{1}{7}\right)$$

$$\times \left(1+\frac{1}{8}\right) \times \left(1-\frac{1}{9}\right)$$

$$= \left(\frac{3}{3}-\frac{1}{3}\right) \times \left(\frac{4}{4}+\frac{1}{4}\right) \times \left(\frac{5}{5}-\frac{1}{5}\right) \times \left(\frac{6}{6}+\frac{1}{6}\right)$$

$$\times \left(\frac{7}{7}-\frac{1}{7}\right) \times \left(\frac{8}{8}+\frac{1}{8}\right) \times \left(\frac{9}{9}-\frac{1}{9}\right),$$

and this obviously simplifies to

$$\left(\frac{3}{3}-\frac{1}{3}\right) \times \left(\frac{4}{4}+\frac{1}{4}\right) \times \left(\frac{5}{5}-\frac{1}{5}\right) \times \left(\frac{6}{6}+\frac{1}{6}\right) \times \left(\frac{7}{7}-\frac{1}{7}\right)$$

$$\times \left(\frac{8}{8}+\frac{1}{8}\right) \times \left(\frac{9}{9}-\frac{1}{9}\right) = \frac{2}{3} \times \frac{5}{4} \times \frac{4}{5} \times \frac{7}{6} \times \frac{6}{7} \times \frac{9}{8} \times \frac{8}{9}.$$

After the first of these factors, we have three pairs of reciprocals that each cancel to 1, and multiplication therefore gives us $\frac{2}{3}$ as the result.

Solution for Problem 7.3:

Similar to the last problem, we can readily calculate

$$\left(1+\frac{1}{2}\right) \times \left(1-\frac{1}{2}\right) \times \left(1+\frac{1}{3}\right) \times \left(1-\frac{1}{3}\right) \times \left(1+\frac{1}{4}\right) \times \left(1-\frac{1}{4}\right)$$

$$\times \left(1+\frac{1}{5}\right) \times \left(1-\frac{1}{5}\right) \times \left(1+\frac{1}{6}\right)$$

$$= \frac{3}{2} \times \frac{1}{2} \times \frac{4}{3} \times \frac{2}{3} \times \frac{5}{4} \times \frac{3}{4} \times \frac{6}{5} \times \frac{4}{5} \times \frac{7}{6},$$

It is now possible to cancel almost all of the numerators, in fact, all but the 7 at the end. What we are then left with in the denominator is a 2, a 5 and several ones, and the result is therefore equal to

$$\frac{7}{2 \times 5} = \frac{7}{10}.$$

Solution for Problem 7.4:

At least we know the year this problem was posed. To actually do the calculation, let's start from the bottom. First of all, we can write

$$\frac{1}{9+\frac{1}{7}} = \frac{7 \times 1}{7 \times \left(9+\frac{1}{7}\right)} = \frac{7}{63+1} = \frac{7}{64}.$$

This gives us

$$1 + \frac{1}{9+\frac{1}{9+\frac{1}{7}}} = 1 + \frac{1}{9+\frac{7}{64}}.$$

Repeating this idea gives us

$$\frac{1}{9+\frac{7}{64}} = \frac{64 \times 1}{64 \times \left(9+\frac{7}{64}\right)} = \frac{64}{576+7} = \frac{64}{583},$$

and therefore

$$1 + \frac{1}{9+\frac{1}{9+\frac{1}{7}}} = 1\frac{64}{583}.$$

Not so easy to spot the year of the problem now, is it?

Solution for Problem 7.5:

As a first step, we note that $\frac{1}{1} = \frac{2}{2} = 1$ obviously holds, and this gives us

$$\frac{6+\frac{1}{2}}{1-\frac{1}{1+\frac{2}{1+\frac{2}{1}}}} = \frac{6+\frac{1}{2}}{1-\frac{1}{1+\frac{2}{1+1}}} = \frac{6+\frac{1}{2}}{1-\frac{1}{1+\frac{2}{2}}} = \frac{6+\frac{1}{2}}{1-\frac{1}{1+1}} = \frac{6+\frac{1}{2}}{1-\frac{1}{2}},$$

which is already a much simpler expression. In order to finish this, we can simply expand the big fraction with the factor 2, which then yields

$$\frac{6+\frac{1}{2}}{1-\frac{1}{1+\frac{2}{1+\frac{2}{1}}}} = \frac{6+\frac{1}{2}}{1-\frac{1}{2}} = \frac{2 \times \left(6+\frac{1}{2}\right)}{2 \times \left(1-\frac{1}{2}\right)} = \frac{12+1}{2-1} = \frac{13}{1} = 13.$$

Solution for Problem 7.6:

Recall that a fraction "of" a number is obtained by multiplication with that fraction. We can therefore write the sentence "Two-thirds of half of $7 \times A$ is equal to 15". As

$$\frac{2}{3} \times \frac{1}{2} \times 7 \times A = 15,$$

and canceling the common factor 2 and multiplying by 3 gives us $7 \times A = 45$. Since $4 \times 7 = 28$, this gives us $28 \times A = 4 \times (7 \times A) = 4 \times 45 = 180$, and since we want to calculate "three-quarters of a third of $28 \times A$", we obtain

$$\frac{3}{4} \times \frac{1}{3} \times 28 \times A = \frac{3}{4} \times \frac{1}{3} \times 180 = 45.$$

Solution for Problem 7.7:

We can write this question in the form of an inequality that will help us find the answer quite readily. The question as posed asks for integers x, such that

$$\frac{1}{4} < \frac{x}{17} < \frac{1}{3}.$$

The answer to this may not seem immediately obvious, but it becomes much easier to see the answer if we write the three fractions in this inequality with a common denominator. Since the numbers 4, 17 and 3 have no common factors, the lowest common denominator is simply their product $4 \times 17 \times 3$, and the inequality can be written as

$$\frac{1 \times 17 \times 3}{4 \times 17 \times 3} < \frac{x \times 4 \times 3}{4 \times 17 \times 3} < \frac{1 \times 4 \times 17}{4 \times 17 \times 3}.$$

Since the denominators are all the same here, we can now compare the numerators and obtain

$$1 \times 17 \times 3 < x \times 4 \times 3 < 1 \times 4 \times 17, \quad \text{or} \quad 51 < 12x < 68.$$

Dividing by 12, this gives us

$$\frac{51}{12} < x < \frac{68}{12},$$

which we can write with mixed numbers as

$$4\frac{1}{4} < x < 5\frac{2}{3}.$$

The only possible integer candidate for x is therefore 5, yielding

$$\frac{1}{4} < \frac{5}{17} < \frac{1}{3}.$$

Solution for Problem 7.8:

Applying the same idea as we used in the last problem, we can write

$$\frac{1}{8} < \frac{m}{50} < \frac{1}{7},$$

which is equivalent to

$$\frac{1 \times 25 \times 7}{8 \times 25 \times 7} < \frac{m \times 4 \times 7}{50 \times 4 \times 7} < \frac{1 \times 8 \times 25}{7 \times 8 \times 25} \quad \text{or} \quad \frac{25 \times 7}{4 \times 7} < m < \frac{8 \times 25}{4 \times 7},$$

which yields

$$\frac{25}{4} < m < \frac{50}{7} \quad \text{or} \quad 6\frac{1}{4} < m < 7\frac{1}{7}.$$

We see that m must be equal to 7, with

$$\frac{1}{8} < \frac{7}{50} < \frac{1}{7}.$$

Solution for Problem 7.9:

If a fraction has a multiple of 6 as its numerator, we can define the numerator as $6x$, with x standing for some positive integer. Since the denominator of this fraction is 5 greater than its numerator, the denominator must then be equal to $6x + 5$. And the fraction can therefore be written as $\frac{6x}{6x+5}$. We are given that the value of this fraction is greater than $\frac{3}{4}$, but less than $\frac{4}{5}$, and this means that we have

$$\frac{3}{4} < \frac{6x}{6x + 5} < \frac{4}{5}.$$

The left-hand inequality is equivalent to $3 \times (6x + 5) < 4 \times 6x$, or $18x + 15 < 24x$, which gives us $15 < 6x$ or $2\frac{1}{2} < x$. Similarly, the right-hand inequality is equivalent to $5 \times 6x < 4 \times (6x + 5)$, or $30x < 24x + 20$, which gives us $6x < 20$ or $x < 3\frac{1}{3}$. Since both of these conditions must hold, we have $x = 3$, and the fraction $\frac{6x}{6x+5}$ is therefore equal to $\frac{6\times 3}{6\times 3+5} = \frac{18}{23}$. The required sum is therefore $18 + 23 = 41$.

Solution for Problem 7.10:

Simply "plugging in" each of the numbers in turn shows us that 6 is the number we are looking for, among the given ones. In turn, we get

$$\frac{4+3}{12} = \frac{9+4}{20} \quad \text{or} \quad \frac{7}{12} = \frac{13}{20},$$

$$\frac{5+3}{12} = \frac{9+5}{20} \quad \text{or} \quad \frac{8}{12} = \frac{14}{20} \quad \text{or} \quad \frac{2}{3} = \frac{7}{10},$$

$$\frac{6+3}{12} = \frac{9+6}{20} \quad \text{or} \quad \frac{9}{12} = \frac{15}{20} \quad \text{or} \quad \frac{3}{4} = \frac{3}{4},$$

$$\frac{7+3}{12} = \frac{9+7}{20} \quad \text{or} \quad \frac{10}{12} = \frac{16}{20} \quad \text{or} \quad \frac{5}{6} = \frac{4}{5}, \quad \text{and}$$

$$\frac{8+3}{12} = \frac{9+8}{20} \quad \text{or} \quad \frac{11}{12} = \frac{17}{20}.$$

It is clear that the only correct statement is the one we obtained by writing 6 instead of the squares.

The more interesting part of this problem is the second question. Since we are perhaps more used to working with letters as variables, let us write x instead of the squares and ask whether there exists another number other than 6 that we can substitute for x in the equation

$$\frac{x+3}{12} = \frac{9+x}{20}$$

to get a true statement. We can multiply this equation by the common denominator 60 of the two fractions, and obtain the equivalent statement

$$60 \times \frac{x+3}{12} = 60 \times \frac{9+x}{20},$$

which is in turn equivalent to $5 \times (x+3) = 3 \times (9+x)$ or $5x+15 = 27+3x$, which gives us $2x = 12$ or $x = 6$. The fact that all of these statements are logically equivalent means that 6 is indeed the only number we can write in place of the squares if we want the resulting expression to be true.

Solution for Problem 7.11:

The only possible combinations of two digits we can choose from among the four given digits are

$$(2, 3), (2, 4), (2, 6), (3, 4), (3, 6) \text{ and } (4, 6).$$

Since each of these pairs must be combined with the pair composed of the other two digits to form fractions of the required type, there are only three possible pairs of such combinations we need to consider, namely

$$(2, 3) \text{ and } (4, 6), (2, 4) \text{ and } (3, 6), \text{ or } (2, 6) \text{ and } (3, 4).$$

We do not need to look at all possible fractions resulting from these combinations, as we know from the suggested results that the numerator must always be smaller than the denominator. This means that we obtain the following possible combinations:

For (2, 3) and (4, 6): $\frac{23}{46} = \frac{1}{2}$, $\frac{23}{64}$, $\frac{32}{46} = \frac{16}{23}$, $\frac{32}{64} = \frac{1}{2}$.

For (2, 4) and (3, 6): $\frac{24}{36} = \frac{2}{3}$, $\frac{24}{63} = \frac{8}{21}$, $\frac{42}{63} = \frac{2}{3}$, $\frac{36}{42} = \frac{6}{7}$.

For (2, 6) and (3, 4): $\frac{26}{34} = \frac{13}{17}$, $\frac{26}{43}$, $\frac{34}{62} = \frac{17}{31}$, $\frac{43}{62}$.

Comparing this complete list of possibilities with the options we were offered, we see that $\frac{1}{2}$, $\frac{2}{3}$, $\frac{6}{7}$ and $\frac{13}{17}$ are indeed all possible results, and only $\frac{5}{6}$ is not.

Solution for Problem 7.12:

Multiplying the given equation

$$\frac{a+b}{a-b} = \frac{7}{4}$$

with the product of the denominators of the two fractions gives us

$$4 \times (a - b) \times \frac{a+b}{a-b} = 4 \times (a - b) \times \frac{7}{4}$$

or $4 \times (a + b) = 7 \times (a - b)$, which is equivalent to $4a + 4b = 7a - 7b$ or $11b = 3a$. This means that $a : b = 11 : 3$, which we can also write as $\frac{a}{b} = \frac{11}{3}$. Squaring gives us the required result as

$$\frac{a^2}{b^2} = \frac{11^2}{3^2} = \frac{121}{9}.$$

Solution for Problem 7.13:

First of all, we can see what happens if we compare two fractions with the same numerator. It is obvious that the fraction with the larger denominator will be the smaller of the two. For instance, compare the fractions $\frac{2}{4}$ and $\frac{2}{5}$. Each of these has the numerator 2. It is obvious that $5 > 4$ holds, but we also have $\frac{2}{4} = 0.5$ and $\frac{2}{5} = 0.4$, and therefore $0.4 < 0.5$ or $\frac{2}{5} < \frac{2}{4}$.

It is not difficult to prove that this will always hold. Let us assume that a is some positive integer and p and q are positive integers with $p < q$. We easily obtain

$$p < q \Rightarrow \frac{1}{q} < \frac{1}{p} \Rightarrow \frac{a}{q} < \frac{a}{p}$$

by some elementary algebraic manipulation.

Now, let us look at the four fractions we are asked to compare. The first three, namely $\frac{5}{x}$, $\frac{5}{x+1}$ and $\frac{5}{x-1}$, all have the same numerator, but $\frac{5}{x+1}$ clearly has the largest denominator, no matter which integer greater than 5 we substitute for x. We therefore know that this expression will assume the smallest value of the three fractions for any given value of x greater than 5. It therefore only remains to compare the value of this fraction to the value of $\frac{x}{5}$.

This is quite easy, though. For any value of x greater than 5, the numerator of $\frac{x}{5}$ is certainly larger than its denominator, which means that the value of the fraction is greater than 1. On the other hand, for any such x, the value of the denominator of $\frac{5}{x+1}$ is greater than the value of its numerator, which means that the value of this fraction is clearly less than 1.

Summing up, we see that the value of the fraction $\frac{5}{x+1}$ is the smallest of the four fractions for any choice of x greater than 5.

Solution for Problem 7.14:

If we let x stand for the numerator of a fraction with the required property, we have

$$\frac{1}{4} < \frac{x}{25} < \frac{1}{2}.$$

It is fairly straightforward to expand this to hundredths as

$$\frac{25}{100} < \frac{4x}{100} < \frac{50}{100},$$

and this means that $25 < 4x < 50$, or $6.25 < x < 12.5$ must hold for the numerators. We see that all values of x from 7 to 12 yield fractions with the required property, and there are therefore 6 such fractions.

Solution for Problem 7.15:

The given equation

$$1 = \frac{1}{2} + \frac{1}{12} + \frac{1}{18} + \frac{1}{A} + \frac{1}{B}$$

is equivalent to

$$\frac{1}{A} + \frac{1}{B} = 1 - \left(\frac{1}{2} + \frac{1}{12} + \frac{1}{18}\right) = 1 - \frac{18 + 3 + 2}{36} = 1 - \frac{23}{36} = \frac{13}{36}.$$

Multiplying by the denominators shows that this is equivalent to

$$\frac{36AB}{A} + \frac{36AB}{B} = \frac{13 \times 36AB}{36} \quad \text{or} \quad 36 \times (A + B) = 13AB.$$

Since 13 is a prime and A and B are positive integers, 36 must be a factor of AB and $A + B$ must be a factor of 13. This means that the values of A and B must be 4 and 9 in some order, and we therefore have $A + B = 4 + 9 = 13$.

Solution for Problem 7.16:

(a) Noting that $3 \times 4 \times 7$ is the lowest common multiple of the denominators, we can multiply by this factor, and the equation is then seen to be equivalent to

$$\frac{3 \times 4 \times 7}{12} + \frac{3 \times 4 \times 7}{28} = \frac{3 \times 4 \times 7}{14} + \frac{3 \times 4 \times 7}{21},$$

and canceling common factors shows that this is equivalent to $7 + 3 = 6 + 4$, which is obviously true.

(b) We can deal with this in a similar way to part (a). The lowest common multiple of the denominators here is $2 \times 3 \times 5 \times 19 \times x$, and multiplying the equation by this factor, we obtain the equivalent equation

$$\frac{2 \times 3 \times 5 \times 19 \times x}{30} + \frac{2 \times 3 \times 5 \times 19 \times x}{95}$$

$$= \frac{2 \times 3 \times 5 \times 19 \times x}{38} + \frac{2 \times 3 \times 5 \times 19 \times x}{x}.$$

Canceling gives us the equivalent equation

$$19x + 6x = 15x + 570,$$

which simplifies to $10x = 570$ or $x = 57$.

(c) and (d) Multiplying a true equation with any number yields another true equation. One easy way to get a different set of integers with the required properties is to start from one of the sets we are given and apply a multiplication of this type. For instance, multiplying the equation from (a) with $\frac{1}{2}$ gives us

$$\frac{1}{12} \times \frac{1}{2} + \frac{1}{28} \times \frac{1}{2} = \frac{1}{14} \times \frac{1}{2} + \frac{1}{21} \times \frac{1}{2} \quad \text{or} \quad \frac{1}{24} + \frac{1}{56} = \frac{1}{28} + \frac{1}{42}.$$

This yields the set $\{24, 56, 28, 42\}$. Since we can do this with any factor of the form $\frac{1}{n}$ (with n being some positive integer), there are obviously infinitely many such sets.

Note that this answers the question as posed, but that there are subtler ways to find such sets. If your curiosity is piqued, this could be the starting point of some interesting research you could do on your own. Can you find another set of four numbers such that the four numbers do not have a common factor greater than 1?

Solution for Problem 7.17:
Let x denote the length of the piece remaining after he wakes up. Since this is half as long as the distance he had slept through, we know that he slept through a piece of length $2x$. Altogether, he fell asleep at the half-way mark of his trip, and this means that the length of each of the pieces before and after he fell asleep was equal to $2x + x = 3x$. The total length of the trip was therefore $3x + 3x = 6x$, and he slept through $\frac{2x}{6x} = \frac{1}{3}$ of it.

Solution for Problem 7.18:
Most people find this type of problem, involving rates and distances, very difficult. Let's see if we can't make the solution at least plausible.

If Klaus walks x meters in t minutes and runs twice as fast as he walks, he runs $2x$ meters in t minutes. His walking speed is $\frac{x}{t}$ meters per minute and his running speed is twice that, namely $\frac{2x}{t}$ meters per minute.

On his way to school, he walks one-third of the 16 minutes, or $\frac{16}{3} = 5\frac{1}{3}$ minutes, and runs twice as long, namely $\frac{32}{3} = 10\frac{2}{3}$ minutes. The total distance he walks on this trip is therefore equal to

$$\frac{x}{t} \times \frac{16}{3} \text{ meters}$$

and the total distance he runs is equal to

$$\frac{2x}{t} \times \frac{32}{3} \text{ meters.}$$

The distance to his school is therefore equal to

$$\frac{x}{t} \times \frac{16}{3} + \frac{2x}{t} \times \frac{32}{3} = \frac{80x}{3t} \text{ meters.}$$

On his way home, we are told that he runs one-third of the time and walks two-thirds of the time. In other words, he walks twice as long as he runs. Since he runs twice as fast as he walks, the distance he walks is equal to the distance he runs, and this is half of the total distance, namely

$$\frac{1}{2} \times \frac{80x}{3t} = \frac{40x}{3t} \text{ meters.}$$

Since he is walking at a speed of $\frac{x}{t}$ meters per minute, it takes him $\frac{40}{3} = 13\frac{1}{3}$ minutes to walk this distance. It takes him half as long, namely $\frac{20}{3} = 6\frac{2}{3}$ minutes to walk this distance, giving us a total time of $13\frac{1}{3} + 6\frac{2}{3} = 20$ minutes for his trip home.

Solution for Problem 7.19:
If $\frac{3}{5}$ of the students in a class are girls, this means that $\frac{2}{5}$ of the students in the class are boys. There are therefore certainly less boys than girls in the class, and if two girls and two boys join the class, this fact cannot change. It is therefore certainly true that there are less boys than girls in the class.

This is enough to answer the question, of course. It is also enough to eliminate the incorrect statements (i) and (ii). But what about statements (iv) and (v)? How can we do a quick check to eliminate these from our considerations?

The easiest way is to consider a small specific number of girls in the class. Of course, if we find that this number makes it possible for the statement to be true, this does not mean that the statement will always be true, but if it is already false for a small number, it cannot be true in general. So, what if there are 3 girls in the class at the beginning? Since this is $\frac{3}{5}$ of the class, there are 5 students in the class at the beginning, and $5 - 3 = 2$ of them are boys. If 2 boys and 2 girls join the class, there are now 5 girls and 4 boys in the class. In this case, $\frac{5}{9}$ of the students in the class are girls, since there are $5 + 4 = 9$ students in the class altogether, and statement (iv) is certainly incorrect.

We have to watch out, though. Just because statement (v) is correct for these specific numbers, does not mean that it will always be correct. So, let us check another number. What if there are initially 6 girls in the class? Since this is $\frac{3}{5}$ of the class, there are 10 students in the class altogether, of which $10 - 6 = 4$ are boys. With 2 boys and 2 girls joining the class, there are then 8 girls and 6 boys, and of all the $8 + 6 = 14$ students in the class, $\frac{8}{14} = \frac{4}{7}$ are girls. We see that statement (v) is not correct in this case, and therefore also not correct in general.

This shows us that statement (iii) is indeed the only one that is certainly correct.

Solution for Problem 7.20:

Between 13:00 and 14:30, Snarlotte goes up $\frac{7}{12} - \frac{1}{3} = \frac{7}{12} - \frac{4}{12} = \frac{3}{12} = \frac{1}{4}$ of the entire distance. Since one and a half hours elapse between 13:00 and 14:30, and one-third of this, namely half an hour, elapses between 13:00 and 13:30, she goes up one-third as much, namely $\frac{1}{3} \times \frac{1}{4} = \frac{1}{12}$ of the total distance in that time. Since she had already gone up $\frac{1}{3}$ of the distance by 13:00, the fraction of the total distance she has climbed up by 13:30 is equal to $\frac{1}{3} + \frac{1}{12} = \frac{4}{12} + \frac{1}{12} = \frac{5}{12}$.

Solution for Problem 7.21:

Right now, the mixture contains $\frac{7}{9}$ of 4.5 liters of water, or $\frac{7}{9} \times 4.5 = \frac{7}{9} \times \frac{9}{2} = \frac{7}{2}$ liters. Adding x liters of water gives us $\frac{7}{2} + x$ liters of water in $4.5 + x$ liters of mixture. Since this is to be eight-ninths of the mixture, we obtain the equation

$$\left(\frac{7}{2} + x \right) \div (4.5 + x) = \frac{8}{9}.$$

We can write this as

$$\frac{\frac{7}{2} + x}{\frac{9}{2} + x} = \frac{8}{9},$$

which is equivalent to

$$\frac{7+2x}{9+2x} = \frac{8}{9} \quad \text{or} \quad 9 \times (7+2x) = 8 \times (9+2x),$$

and this simplifies to $63 + 18x = 72 + 16x$, or $2x = 9$, which gives us $x = 4.5$. The mixture will therefore then contain a total of $4.5 + 4.5 = 9$ liters, of which $3.5 + 4.5 = 8$ liters are water.

Solution for Problem 7.22:
This is a pretty straightforward problem, no matter how you tackle it. Perhaps the easiest is simply to consider how much apple juice is left at the end of each day. He drinks half of whatever mixture is in the bottle each day, and it means that he drinks half of the remaining apple juice each day. After the first day, $\frac{1}{2}$ liter of juice is left. After the second day, there is still half of that, i.e. $\frac{1}{4}$ liter. After the third day, half of that is left, that is $\frac{1}{8}$ liter, and after the fourth day, half of that is left, that is $\frac{1}{16}$ liter. We see that he consumed $1 - \frac{1}{16} = \frac{15}{16}$ of a liter of juice in those first four days, leaving a pretty watery drink in the bottle.

Solution for Problem 7.23:
Let x denote the number of cherries in the bowl at the beginning. We are told that Achmed first eats 75% of them, and this means that he leaves 25%, or $\frac{1}{4}$ of them in the bowl. He then eats $\frac{5}{6}$ of the remaining cherries, which means that he leaves $\frac{1}{6}$ of them. The four cherries left over at the end are therefore $\frac{1}{6}$ of $\frac{1}{4}$ of the original cherries, and this gives us the equation

$$\frac{1}{6} \times \frac{1}{4} \times x = 4.$$

Multiplying by the denominators, we see that there were originally $x = 6 \times 4 \times 4 = 96$ cherries in the bowl.

Solution for Problem 7.24:
Let x denote the total value of the mathematician's estate. Claudia is to receive €$600 + \frac{1}{3} \times (x - €600)$. Franz is to receive €$950 + \frac{1}{4} \times x$, and Katharina is to receive $\frac{1}{6} \times x$. Since the total of these three bequests is the

total estate value, we obtain the equation

$$€\,600 + \frac{1}{3} \times (x - €\,600) + €\,950 + \frac{1}{4} \times x + \frac{1}{6} \times x = x.$$

The can be written as

$$€\,600 - \frac{1}{3} \times €\,600 + \frac{1}{3} \times x + €\,950 + \frac{1}{4} \times x + \frac{1}{6} \times x = x$$

or

$$€\,600 - €\,200 + €\,950 = x - \frac{1}{3} \times x - \frac{1}{6} \times x - \frac{1}{4} \times x,$$

which simplifies to $€\,1350 = \frac{1}{4} \times x$. Multiplying by 4, we see that the total value of the estate is equal to $4 \times €\,1350 = €\,5400$.

It is now quite straightforward to calculate the individual bequests. Claudia receives

$$€\,600 + \frac{1}{3} \times (€\,5400 - €\,600) = €\,2200,$$

while Franz receives

$$€\,950 + \frac{1}{4} \times €\,5400 = €\,2300$$

and Katharina receives

$$\frac{1}{6} \times €\,5400 = €\,900.$$

Solution for Problem 7.25:

Let x denote the number of questions posed on the test. Since 9 of the first 10 and $\frac{3}{10}$ of the remaining $x - 10$ are half of all the questions, we have

$$9 + \frac{3}{10} \times (x - 10) = \frac{1}{2} \times x.$$

This simplifies to

$$9 + \frac{3}{10} \times x - \frac{3}{10} \times 10 = \frac{1}{2} \times x$$

or

$$9 - 3 = \frac{5}{10} \times x - \frac{3}{10} \times x,$$

which yields $6 = \frac{2}{10} \times x$ or $x = 30$. The number of correct answers is therefore equal to 9 plus $\frac{3}{10}$ of the remaining 20, for a total of $9 + 6 = 15$, which is indeed half of 30.

Chapter 8

Percentages

Number relationships are often expressed with percentages. Since this particular mathematical tool is so common, you might think that most everyone would be able to interpret number expressions expressed with it quite easily. This is not always the case, however. In fact, many people find correct manipulation of percentages to be quite a tricky business. Here are some problems involving percentages, starting with very elementary one-step problems, and moving slowly to slightly more involved questions.

PROBLEMS

Problem 8.1:
Which of the following amounts is the largest?
(A) 17% of € 1200   (B) 50% of € 400   (C) 41% of € 500   (D) 26% of € 800
(E) 23% of € 900

(StU-07-A3)

Problem 8.2:
8 is 25% of A. B is 16% of 500. How many percent of B is A?

(S5K-14-5)

Problem 8.3:
The number 18 is 40% of a certain number x. The number x is 60% of a number y. Determine y.

(S5K-12-4)

Problem 8.4:

15% of a number is 27. How much is 25% of that number?

(S5K-09-4)

Problem 8.5:

10% of a is b. b% of c is a. Determine the value of c.

(StU-15-A4)

Problem 8.6:

The price of a lift pass in the skiing region of Amapecunia was raised by 10% in each of four successive years. By how many percent was the price raised in that time altogether?

(StU-13-A3)

Problem 8.7:

There are 250 students in the Altenmarkt school. 24% of them like to play soccer and 15% of these play for the local team. How many of the students in the school play for the local team?

(S5K-17-2)

Problem 8.8:

A 4 kg watermelon is composed of 99% water. After the watermelon has been in the sun for a while, it dries up quite a bit and then is only composed of 98% water. How heavy is the watermelon then?

(StU-94-2)

Problem 8.9:

Ten years ago, George spent 30% of his income on rent. Since then, his income has risen by 80%, but his rent has only gone up by 50%. What percentage of his income does he currently spend on his rent?

(S5K-02-7)

Problem 8.10:

There are a number of bunnies in a coop, each of which is either black or white. This year, twice as many black bunnies were added to the population as white bunnies were added. Because of this, the percentage of black bunnies in the coop rose from 40% to 50%. How many percent more bunnies are in the coop this year than were there last year?

(S5K-99-7)

Problem 8.11:
A number of parrots live in the Parowald forest. Last week, 70% of the parrots were blue. Half of the blue parrots flew away and all others stayed in the forest. In the meantime, $\frac{3}{7}$ of the parrots that had left have returned. What percentage of the parrots in the forest are blue now?

(S5K-15-5)

Problem 8.12:
The Sorghuber sausage factory makes a special kind of baloney with a fat content of 40%. The sausage-meister would like to create a new kind of diet baloney using a similar recipe. For this purpose, he plans to use half as much fat per portion as he would in the regular recipe, but the same amount of every other ingredient. What will be the fat content of the diet baloney?

(StU-14-A7)

Problem 8.13:
Among the female students in a school, 18% wear glasses. Among the boys there, 24% do. What percentage of all students wears glasses, if we know that $\frac{9}{20}$ of the students in the school are female?

(S5K-98-8)

Problem 8.14:
The students in the 5a class were supposed to do their homework. 60% of them did not do their German homework, 80% didn't do their English homework, and 75% didn't do their Math homework. At least how many percent didn't do any homework at all?

(S5K-04-6)

Problem 8.15:
There are 140 gummi bears in a jar. 30% of these are pineapple-flavored and 70% are papaya. Remi doesn't like pineapple, and only eats some of the papaya ones. After a while he counts them and notes that only 30% of

the gummi bears in the jar are now papaya. How many gummi bears are in the jar now?

(S5K-04-9)

TIPS

Tip for Problem 8.1:
This is a matter of simple calculation, of course. It is easier to do in your head if you note that a percentage is the same as hundredths, and one-hundredth of a number with zeros at the end is obtained by simply dropping the last two of these zeros.

Tip for Problems 8.2, 8.3 and 8.4:
For these three problems, it is good to recall that writing $n\%$ is the same as writing $\frac{n}{100}$. Rewriting the problem statements as equations and using these fractions will make it easy to find the answers.

Tip for Problem 8.5:
This is slightly more abstract than the last few problems, but really pretty much the same, if you think about it. You can certainly apply exactly the same tools to solve it.

Tip for Problem 8.6:
You can write a percentage increase as a multiple of the original cost.

Tip for Problem 8.7:
This one is pretty much just a straightforward calculation. How many play soccer? Once you have calculated that, you are practically done.

Tip for Problem 8.8:
If you have never seen a problem like this, you will almost certainly be surprised by the answer! How much of the watermelon isn't composed of water? That part doesn't evaporate in the sun, so it won't change.

Tip for Problem 8.9:
Once again, the introduction of a variable to stand for an unknown quantity will help a lot.

Tip for Problem 8.10:
There are two quantities we don't know here, namely the number of bunnies in the coop at the beginning and the number of bunnies that are added to the coop. Try defining a variable for each of these quantities and then writing what you know in algebraic notation.

Tip for Problem 8.11:
Once again, introduce a variable for an important quantity you don't know the value of. In this case, the original number of parrots in the forest looks like a good candidate.

Tip for Problem 8.12:
This is a subtle one. It would seem that the answer should be 20%, right? But watch out! If he uses half as much fat as usual, he is also producing less sausage with the same amount of lean. Once again, it will be a good idea to introduce a variable to stand for some quantity you don't know the value of, like the weight of a standard sausage.

Tip for Problem 8.13:
This time it will prove useful to define a variable to describe the total number of students in the school.

Tip for Problem 8.14:
This problem is quite different from the others in this section! We just need some basic logic to solve this one. Oh, and we need to know that all of the students make up 100% of the class, of course.

Tip for Problem 8.15:
We know the total number of gummi bears, so there is no point in defining that as a variable. On the other hand, we do not know how many gummi bears Remi ate, so that makes this number a good candidate for a variable.

SOLUTIONS

Solution for Problem 8.1:
Since we can calculate percentages of a number by simply dropping the last two zeros, we see that 17% of € 1200 is equal to $17 \times 12 = $ € 204. Similarly, we obtain 50% of € 400 $= 50 \times 4 = $ € 200, 41% of € 500 $= 41 \times 5 = $ € 205,

26% of € 800 = 26 × 8 = € 208 and 23% of € 900 = 23 × 9 = € 207. The largest of these amounts is therefore 26% of € 800 = € 208.

Solution for Problem 8.2:
We are given $8 = \frac{25}{100} \times A$ and $B = \frac{16}{100} \times 500$. This means that $A = 4 \times 8 = 32$ and $B = 16 \times 5 = 80$. In order to calculate the required percentage, we therefore obtain

$$\frac{32}{80} \times 100 = 40.$$

A is therefore 40% of B.

Solution for Problem 8.3:
We are given $18 = \frac{40}{100} \times x$ and $x = \frac{60}{100} \times y$. From the first equation, we obtain

$$x = 18 \times \frac{100}{40} = 45,$$

and from the second, we therefore obtain

$$y = x \times \frac{100}{60} = 45 \times \frac{5}{3} = 75,$$

which completes the calculation.

Solution for Problem 8.4:
If we let x denote the unknown number in question, we are given

$$\frac{15}{100} \times x = 27.$$

This means that we can calculate the value of x as

$$x = 27 \times \frac{100}{15} = 180,$$

and we therefore obtain 25% of this number as

$$\frac{25}{100} \times 180 = 45.$$

Solution for Problem 8.5:

We are given $\frac{10}{100} \times a = b$ and $\frac{b}{100} \times c = a$. The first equation can also be written as $a = 10b$, and plugging this value into the second equation gives us

$$\frac{b}{100} \times c = 10b,$$

which yields $c = 1000$.

Solution for Problem 8.6:

If a number x is increased by 10%, the result is equal to $x + \frac{10}{100} \times x = 1.1 \times x$. Since this happens four times in a row, the resulting price is equal to

$$1.1 \times (1.1 \times (1.1 \times (1.1 \times x))) = 1.1^4 \times x = 1.4641 \times x.$$

This means that the original cost of x was increased by $0.4641 \times x$, or 46.41% of x.

Solution for Problem 8.7:

If 24% of the 250 students play soccer, that makes a total of $\frac{24}{100} \times 250 = 60$ soccer players. Of these, 15% play for the team, and that makes a total of $\frac{15}{100} \times 60 = 9$ students from the school, who play for the local team.

Solution for Problem 8.8:

If 99% of a watermelon is water, we know that 1% of it isn't. This means that 1% of 4 kg, or $\frac{1}{100} \times 4 = \frac{1}{25}$ of a kg of the watermelon is not composed of water. This does not change in the sum, and after evaporation of some of the water in the melon, this $\frac{1}{25}$ of a kg is (100–98)%, or 2% of the mass of the melon. If we let x denote the mass of the melon after evaporation, we therefore obtain

$$\frac{1}{25} = \frac{2}{100} \times x,$$

and multiplying by 50 gives us $x = 2$ kg. The mass of the watermelon after it has been in the sun is therefore only 2 kg, or half of what it was before!

Solution for Problem 8.9:

Let x denote George's original income. Since he spent 30% of x on rent, his rent at the time was $\frac{30}{100} \times x = \frac{3x}{10}$.

Now, his income has risen by 80%, and his current income is therefore equal to $x + \frac{80}{100} \times x = 1.8 \times x$. On the other hand, his rent has gone up by 50%, and his current rent is therefore equal to $\frac{3x}{10} + \frac{50}{100} \times \frac{3x}{10} = 1.5 \times \frac{3x}{10} = \frac{3}{2} \times \frac{3x}{10} = \frac{9x}{20}$. We wish to know what percentage of his current income this amounts to. The percentage is given by the expression

$$\frac{\frac{9x}{20}}{1.8 \times x} \times 100 = \frac{9x \times 5}{1.8 \times x} = 25.$$

We see that George currently spends 25% of his income on rent.

Solution for Problem 8.10:
Let x denote the number of bunnies in the coop before the new group is added, and let y denote the number of white bunnies that are added to the population. Since we know that twice as many black bunnies as white bunnies were added to the population, the number of black bunnies that were added is therefore equal to $2y$, and the total number of bunnies in the coop after the new ones were added is therefore equal to $x + y + 2y = x + 3y$.

We know that 40% of the original bunnies were black, and this number is equal to $\frac{40}{100} \times x = \frac{2x}{5}$. After the new bunnies were added, we know that 50% are black, and this number is equal to

$$\frac{50}{100} \times (x + 3y) = \frac{x + 3y}{2}.$$

We know that this number resulted from the original number of black bunnies in the coop by adding $2y$ bunnies, and this gives us

$$\frac{x + 3y}{2} = \frac{2x}{5} + 2y.$$

Multiplying by 10, we see that this is equivalent to

$$5x + 15y = 4x + 20y \quad \text{or} \quad x = 5y,$$

which we can also write as $y = \frac{x}{5}$. Since $3y$ bunnies were added to the original x bunnies in the coop, the number of bunnies went up by $\frac{3x}{5} = \frac{60}{100} \times x$, and we see that there are 60% more bunnies in the coop this year than there were last year.

Solution for Problem 8.11:
Let x denote the number of parrots originally in the forest. Since 70% of them were blue, the number of blue parrots originally in the forest was equal

to $\frac{70}{100} \times x = \frac{7x}{10}$. We know that half of these flew away, and their number as well as the number of the parrots staying is therefore equal to $\frac{1}{2} \times \frac{7x}{10} = \frac{7x}{20}$. Next, we are told that $\frac{3}{7}$ of these have returned, and the number of blue parrots in the forest now is therefore equal to

$$\frac{7x}{20} + \frac{3}{7} \times \frac{7x}{20} = \frac{7x}{20} + \frac{3x}{20} = \frac{10x}{20} = \frac{x}{2}.$$

Of course, this does not mean that half of the parrots currently in the forest are blue, since there are now less parrots in the forest than there were originally. We know that $\frac{7x}{20}$ parrots left and $\frac{3}{7} \times \frac{7x}{20} = \frac{3x}{20}$ parrots returned. This means that the current total number of parrots in the forest is equal to

$$x - \frac{7x}{20} + \frac{3x}{20} = \frac{16x}{20} = \frac{4x}{5},$$

and the percentage of blue parrots currently in the forest is therefore equal to

$$\frac{\frac{x}{2}}{\frac{4x}{5}} \times 100 = \frac{5x}{8x} \times 100 = 62.5.$$

We see that 62.5% of the parrots currently in the forest are blue.

Solution for Problem 8.12:
Let x denote the mass of a standard baloney, in grams. We know that 40% of this, or

$$\frac{40}{100} \times x = \frac{2x}{5} \text{ g,}$$

is fat, while 60% of this, or

$$\frac{60}{100} \times x = \frac{3x}{5} \text{ g,}$$

is not. If the sausage-meister uses half as much fat for the diet baloney but leaves the rest of the recipe unchanged, this means that he is using

$$\frac{1}{2} \times \frac{2x}{5} = \frac{x}{5} \text{ g}$$

of fat, along with $\frac{3x}{5}$ g of other ingredients, for a total of $\frac{x}{5} + \frac{3x}{5} = \frac{4x}{5}$ g. The fat content of the diet baloney is then therefore equal to

$$\frac{\frac{x}{5}}{\frac{4x}{5}} \times 100 = \frac{x}{4x} \times 100 = 25\%.$$

Solution for Problem 8.13:

If we let x denote the number of students in the school, this will just be a straightforward calculation. Since $\frac{9}{20}$ of the students in the school are female, the other $1 - \frac{9}{20} = \frac{11}{20}$ are assumed to be boys. (For the highly limited purposes of this problem, we must assume that all students identify as either boys or females, as we would otherwise not have enough information given to solve it.) Since 18% of the female students wear glasses, the number of female students who wear glasses is equal to

$$\frac{18}{100} \times \frac{9}{20} \times x = \frac{81x}{1000},$$

and since 24% of the boys wear glasses, the number of boys who wear glasses is equal to

$$\frac{24}{100} \times \frac{11}{20} \times x = \frac{132x}{1000}.$$

Altogether, there are therefore $\frac{81x}{1000} + \frac{132x}{1000} = \frac{213x}{1000}$ such students, and this means that 21.3% of the students in the school wear glasses.

Solution for Problem 8.14:

First of all, let us ignore the math homework for the moment. (We know that 75% of the class didn't!) Just considering German and English, we note that $60\% + 80\% = 140\%$, which means that at least 40% of the students in the 5a class did not do either their German homework or their English homework. (It is possible that 40% did their German homework and 20% did their English homework, but nobody did both. That leaves $100\% - 40\% - 20\% = 40\%$ who did neither.)

Now, we can bring Math back into consideration. Since at least 40% of the students did neither of their language assignments, and $40\% + 75\% = 115\%$, we see that there must have been at least 15% that didn't do any homework at all. (It is possible that all of the students that did some homework only completed their assignments in one subject; 40% in German, 20% in English and 25% in Math, leaving $100\% - 40\% - 20\% - 25\% = 15\%$ who did nothing.)

Solution for Problem 8.15:

Let x denote the number of gummi bears Remi ate. We know that these were all papaya-flavored. Of the original 140 gummi bears, 70% were papaya-flavored, and there were therefore $\frac{70}{100} \times 140 = 98$ of these. After Remi ate

x of them, there were $98 - x$ papaya-flavored gummi bears left, out of a total of $140 - x$ gummi bears. Since we know that these were 30% of the total number remaining, we have

$$\frac{98 - x}{140 - x} \times 100 = 30 \quad \text{or} \quad 10 \times (98 - x) = 3 \times (140 - x).$$

This simplifies to $980 - 10x = 420 - 3x$ or $560 = 7x$, which yields $x = 80$. We see that Remi ate 80 gummi bears out of the original 140, leaving $140 - 80 = 60$ gummi bears in the jar.

Note that this does solve the problem, as there would have been $98 - 80 = 18$ papaya gummi bears left out of 60, and since $\frac{18}{60} \times 100 = 30$, it is true that 30% of the remaining gummi bears are now papaya-flavored.

Chapter 9

Our Three-Dimensional World

We do live in an essentially three-dimensional (3D) world, but as was already noted in Chapter 3, many people find 3D problems to be among the most difficult to solve. You may want to try your hand at the problems there before you attack these slightly more general ones. Most of the problems in this section can be done with pure spatial reasoning, though, maybe combined with a little bit of simple calculation. A lot of them have pictures, and if you find the situation difficult to comprehend at first, you should definitely take a closer look at the picture.

PROBLEMS

Cuboids and Volume

Problem 9.1:
A cube-shaped canister has edges 24 cm in length. It is half filled with water. The water is then emptied out (without spillage) into a container in the shape of a rectangular parallelepiped, which is 27 cm wide, 16 cm deep, and 20 cm high. How much from the top is the water level in the new container?

(S5K-98-1)

Problem 9.2:
The two solids in the figure have the same volume. One of the solids is a cube and the other is a cuboid. Both have integer edge lengths. How large is x?

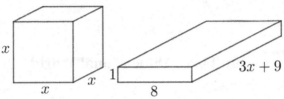

(S5K-02-10)

Problem 9.3:
The base of a cuboid has a perimeter of 46 cm. The perimeters of the right face and the frontal face are 66 cm and 80 cm, respectively. Determine the volume of the cuboid.

(StU-08-II-5)

Problem 9.4:
The block shown in the figure has a surface area of 270 cm². Determine its volume. (We know that all the surfaces except the L-shaped ones are rectangles.)

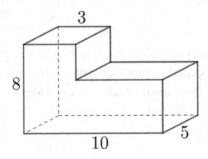

(StU-98-B2)

Problem 9.5:

A wooden cube *ABCDEFGH* has the surface area 54 cm². It is cut through twice with a saw. One of the plane cuts passes through points *B*, *D* and *E* and the other through points *C*, *F* and *H*. Determine the volume of the piece of the cube between the two cuts.

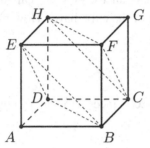

(StU-15-B5)

Problem 9.6:

In the figure we see what remains from a solid cube after we have cut off a corner in such a way that the corners of the resulting triangle are the midpoints of the edges of the original cube. If we cut off all of the corners of the cube in this way, a solid remains with squares and triangles as faces. How many squares and how many triangles make up the surface of this object? Determine its volume if the edges of the original cube are all 6 cm in length.

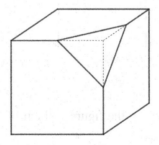

(StU-95-11)

Edges and Such

Problem 9.7:

Two edges of a cube are called "friends" if they have a corner of the cube in common. They are called "acquaintances" if they are parallel. If they are neither friends nor acquaintances, they are called "enemies". How many enemies does each edge of the cube have? How many common friends do two enemies have?

(O-98-2)

Problem 9.8:

A prism has of a total of 2008 faces. How many edges does the prism have?

(StU-08-A5)

Problem 9.9:

We are given a cube with edge length 3 cm. Each edge is divided into three sections of equal length by two points. After this step, each of the 8 corners is cut off in such a way that the plane of the cut joins three of these points, as shown in the figure, each of which is 1 cm from a common corner. How many edges does the resulting solid have?

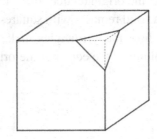

(S5K-03-9)

Problem 9.10:

Consider the cube shown in the figure. Which of the following claims is false?

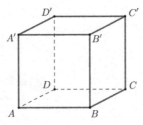

(a) $A'C \perp BD$ (b) $AC \perp B'D'$ (c) $AD' \perp D'C$ (d) $AD' \perp B'C$
(e) $D'C \perp BC$

(StU-93-2)

Problem 9.11:

In the figure, we see a regular, 6-sided prism. This prism has 12 vertices. If we draw the line segment joining two of these vertices, the segment can either lie on the surface of the prism, as is the case for the edge FL or the surface diagonal BI in the figure, or it can be contained in the interior of the prism, as is the case for AK in the figure. How many such line segments lie in the interior of the prism?

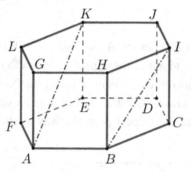

(StU-04-A9)

Some Spatial Reasoning Required (But Not Much)

Problem 9.12:

The famous photographer Annie Liebescherz has chosen the theme "Geometry" for a series of works. She takes individual pictures of the following objects:

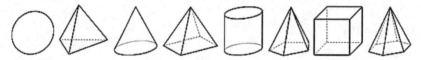

On one of her artistically valuable but fuzzy and vague photos, she sees an object with a hexagonal outline. Unfortunately, she no longer remembers, which object it was. How many of the objects in the picture could it have been?

(StU-13-A6)

Problem 9.13:
A geometric object appears as a square when seen both from the front and from the top. (i.e. the front view and the plan view). Which of the following cannot be the object in question?

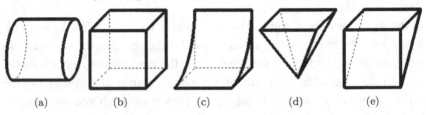

(a)   (b)   (c)   (d)   (e)

(StU-14-A10)

Problem 9.14:
A spider is sitting in the corner A of a house with the total height 10 m and measurements as given in the figure. It wants to crawl to point E, but because of an evil lurking cat, it cannot crawl on the sides of the house. It therefore decides to crawl over the roof along the shortest possible route from A to E, crossing the ridge PQ on the way as shown. How long is the spider's route?

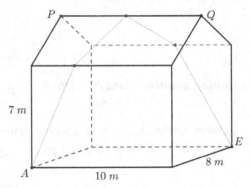

(StU-15-A10)

TIPS

Tip for Problem 9.1:
As a first step, calculate half the volume of the cube. This volume does not change when you pour the water into the other container.

Tip for Problem 9.2:
If the volumes of the two objects are equal, it shouldn't be too hard to set up an equation for x.

Tip for Problem 9.3:
Introduce a, b and c as the values of length, width and height of the cuboid, respectively.

Tip for Problem 9.4:
You can either think of this solid as a cuboid, from which a smaller cuboid has been removed on the upper right-hand side, or as the union of two smaller cuboids (left–right or top–bottom). Once you have decided on one of these interpretations, determine the missing measurements of the cuboids, and you are almost done.

Tip for Problem 9.5:
Each cut saws off a special kind of pyramid, and it shouldn't be too hard to calculate the volume of such a pyramid. First, you'll need to find the edge-length of the cube, though.

Tip for Problem 9.6:
This one is very similar to the last one. Once again, we are sawing special pyramids off of a cube. This time, we already know the edge-length of the cube, though.

Tip for Problem 9.7:
It shouldn't be too hard to find all of the friends of a cube's edge. The same is true of its acquaintances. All of the other edges are enemies.

Tip for Problem 9.8:
If the surface of a prism has 2008 faces, two of its faces must be 2006-gons.

Tip for Problem 9.9:
The edges of the cube remain edges of the resulting solid, although they become shorter. New edges result from cutting off the vertices of the cube.

Tip for Problem 9.10:
The angle between two skew lines is defined as the angle between the parallels to these lines through a common point. What happens when you shift one of the segments in each of the pairs to a parallel position through one of the ends (or the midpoint) of the other?

Tip for Problem 9.11:

If a diagonal lies in the interior of the prism, it must join a vertex of the bottom hexagon with a vertex of the top hexagon.

Tip for Problem 9.12:

This one will take some imagination. Try turning the objects around in your head. What possible outlines can they have in their different positions? Start with the sphere; that one should be easy. What about the other round objects, like the cone? What about the tetrahedron (the triangular pyramid)?

Tip for Problem 9.13:

It is enough to consider the front view of each of these objects.

Tip for Problem 9.14:

The shortest route between two points in a plane is a straight line. Maybe you can find a way to transfer the spider's path into a plane without changing its length.

SOLUTIONS

Solution for Problem 9.1:

Half the volume of the cube is equal to $24 \times 24 \times 12 \, \text{cm}^3$. When this is poured into the other container, the base of which has the measurements $27 \, \text{cm} \times 16 \, \text{cm}$, it forms a cuboid with the altitude a, and therefore with the volume $27 \times 16 \times a \, \text{cm}^3$. Since the volume of the water has not changed, we obtain $27 \times 16 \times a = 24 \times 24 \times 12$, or

$$a = \frac{24 \times 24 \times 12}{27 \times 16} = \frac{2^3 \times 3 \times 2^3 \times 3 \times 2^2 \times 3}{3^3 \times 2^4} = 2^4 = 16 \, \text{cm}.$$

Since the water is 16 cm high in the cuboid, the water level is $20 - 16 = 4$ cm under the top.

Solution for Problem 9.2:

The volume of the cube is x^3, and the volume of the cuboid is $1 \times 8 \times (3x+9)$. Since these two volumes are equal, we obtain the equation

$$x^3 = 8 \times (3x + 9),$$

which is equivalent to

$$x^3 - 24x - 72 = 0.$$

In order to find a value of x that solves this equation, we can insert the values 1, 2, 3 and so on for x, hoping for a true statement at some point. It is simple enough to check that none of the values from 1 through 5 work, but inserting $x = 6$ yields

$$6^3 - 24 \times 6 - 72 = 216 - 144 - 72 = 0,$$

and we see that $x = 6$ is a solution to the problem.

This is enough to answer the question as posed, but a curious reader might want to know whether this is the only possible solution. After all, a general cubic equation (which is exactly what $x^3 - 24x - 72 = 0$ is) can have as many as three different solutions, and it could therefore well be the case that 6 is not the only positive number solving the equation.

Since 6 is a solution, we know that the linear polynomial $x - 6$ must be a divisor of the cubic polynomial $x^3 - 24x - 72$, and we can therefore calculate

$$(x^3 - 24x - 72) \div (x - 6) = x^2 + 6x + 12,$$

which yields

$$x^3 - 24x - 72 = (x - 6)(x^2 + 6x + 12).$$

This means that any positive solution of the quadratic equation

$$x^2 + 6x + 12 = 0$$

is also a possible value of x. This is not possible, however, since the value of

$$x^2 + 6x + 12 = (x^2 + 6x + 9) + 3 = (x + 3)^2 + 3$$

can never be less than 3, no matter what value of x is used, and therefore never equal to 0. The only possible value of x is therefore 6.

Solution for Problem 9.3:
If we let a, b and c denote the values of length, width and height of the cuboid, we can express the perimeters of the base, the right face and the frontal face as $2(a + b)$, $2(b + c)$ and $2(c + a)$, respectively. Since these

perimeters are given as 46 cm, 66 cm and 80 cm, respectively, we obtain the system of equations

$$2(a + b) = 46, \quad 2(b + c) = 66, \quad 2(c + a) = 80,$$

which simplifies to

$$a + b = 23, \quad b + c = 33, \quad c + a = 40.$$

Adding all three equations gives us

$$2(a + b + c) = 96 \quad \text{or} \quad a + b + c = 48,$$

and subtracting the three equations of the simplified system of equations from this equation gives us the values

$$a = 15, \quad b = 8, \quad c = 25$$

for the sides of the cuboid. It is now straightforward to calculate its volume as

$$V = abc = 15 \times 8 \times 25 = 3000 \, \text{cm}^3.$$

Solution for Problem 9.4:
The block can be extended to a cuboid by adding on a smaller cuboid in the upper right as shown in the following figure.

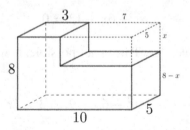

The small cuboid is obviously 5 cm deep. It is 7 cm wide, as its width is given by $10 - 3 = 7$. We do not, however, know its height, and we can therefore name it x.

The volume of the given block is equal to the volume of the large resulting cuboid minus the volume of this small cuboid. The volume of the large cuboid is given as $10 \times 5 \times 8 = 400 \, \text{cm}^3$, and the volume of the small cuboid is equal to $7 \times 5 \times x = 35x$, and the volume of the block is therefore equal to

$$400 - 35x \text{ cm}^3.$$

If we can determine the value of x, we can therefore determine the volume of the block. In order to do this, we recall that the surface area of the block was given as being equal to 270 cm^2. The surface is made up of rectangles on the left and right, on the bottom and top, and L-shaped hexagons in the front and back.

The area of the bottom rectangle is equal to $10 \times 5 = 50 \text{ cm}^2$, and this is also the sum of the areas of the two top rectangles, independent of the value of x, as we can imagine pushing the lower of these rectangles up to the height of the top one, yielding a rectangle the same size as the bottom one. Similarly, the rectangle on the left has the area $8 \times 5 = 40 \text{ cm}^2$, and this is also the sum of the areas of the two rectangles on the right, again independent of the value of x. The area of each of the L-shaped faces is equal to the area of the large front rectangle minus the area of the small front rectangle, i.e. $10 \times 8 - 7 \times x = 80 - 7x$. This means that the total surface area of the block is equal to

$$2 \times 50 + 2 \times 40 + 2 \times (80 - 7x) = 340 - 14x,$$

and since this surface area was given as being equal to 270 cm^2, we obtain the equation

$$340 - 14x = 270,$$

which is equivalent to $14x = 70$ or $x = 5$.

The volume of the block is therefore equal to

$$400 - 35 \times 5 = 400 - 175 = 225 \text{ cm}^3.$$

Note that there are other ways to obtain this result. For instance, we obtain similar expressions (and the same end-result, of course), if we think of the block as being made up of two cuboids that have been glued together, rather than the difference of two cuboids. The two ways to do this are shown in the following figure:

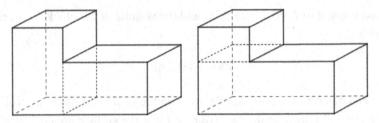

You might enjoy figuring out the intermediate steps using one of these ideas, and comparing the method to the one presented in its entirety here.

Solution for Problem 9.5:

If the edges of the cube have the length a, the surface area is equal to $6a^2$. Since the surface area of the cube is given as $54 \, \text{cm}^2$, we have

$$6a^2 = 54,$$

which yields $a^2 = 9$ or $a = 3$.

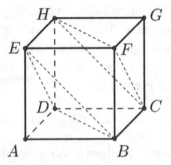

This means that the volume of the cube is equal to $3^3 = 27 \, \text{cm}^3$. In order to find the volume we are asked to determine, we must subtract the volume of the two pyramids that are cut off by the planes BDE and CFH.

Each of these has a right isosceles triangular base with sides of length a, and altitude a. The volume of each of these pyramids is therefore equal to

$$\frac{1}{3} \times 3 \times \frac{1}{2} \times 3^2 = \frac{9}{2},$$

and the volume of the piece of the cube we are searching for is therefore equal to

$$27 - 2 \times \frac{9}{2} = 18 \, \text{cm}^3.$$

Solution for Problem 9.6:

In the figure below, we see the steps leading to the described result. In the middle picture, the top front right-hand corner has been removed, and in the right-most picture, all corners have been removed. We see that the resulting object has a triangular face where each of the vertices of the original cube was, and since the cube had eight vertices, the new object has eight triangular faces. Furthermore, a square remains where each of the original square faces of the cube was, the corners of which are the midpoints of the sides of the faces of the cube. This means that there are as many square faces on the new object as there were on the cube, i.e. six.

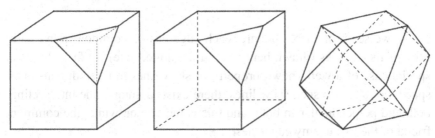

Now, to calculate the volume of the resulting object (the name of which is "cubo-octahedron", by the way) if we are given that the edges of the cube are each 6 cm long, we first note that the volume of the cube is equal to $6^3 = 216\,\text{cm}^3$. In order to obtain the cubo-octahedron, we cut off eight right pyramids (as was the case in the previous problem), each of which has perpendicular edges of length 3 cm. the volume of each such a pyramid is therefore

$$\frac{1}{3} \times 3 \times \frac{1}{2} \times 3^2 = \frac{9}{2},$$

and the volume of the cubo-octahedron is therefore equal to

$$216 - 8 \times \frac{9}{2} = 180\,\text{cm}^3.$$

Solution for Problem 9.7:

A cube has 12 edges. In the left picture below, we see a vertical thick edge, whose four "friends" (two emanating from each of the two end vertices of the edge; top and bottom) are drawn as dashed lines. The edge also has three parallel "acquaintances", drawn as thin full lines, as the 12 edges can be grouped into three groups of four parallel edges. It follows that four of the 12 edges are enemies of the thick edge, since we have $12 - 1 - 4 - 3 = 4$.

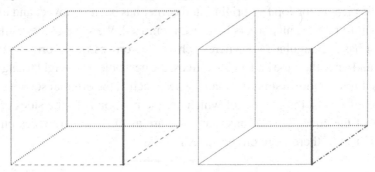

As we see in the right picture, each edge has on common friend with each of its enemies (drawn here as a stroke-dotted line). In fact, this is a special case of a well-known property of skew lines in three-dimensional space. For any two such skew lines, there exists a unique line intersecting both and perpendicular to both, and this is the line on which the common friend of the two enemy edges lies.

Solution for Problem 9.8:

Each of the two 2006-gons on the surface of the prism, i.e. its bases, has 2006 edges. Since there are 2006 more edges connecting a vertex of the first 2006-gon to a vertex of the second 2006-gon, the prism has a total of $3 \times 2006 = 6018$ edges.

Solution for Problem 9.9:

There are several ways to solve this problem. One way is to note that the 12 edges of the cube remain edges of the resulting solid, and that three new edges are formed at each of the 8 vertices of the cube by cutting the little pyramid off. This means that the total number of edges of the resulting solid is equal to

$$12 + 3 \times 8 = 36.$$

Another way to find the answer is to consider what happens with each of the square faces of the cube. Since a triangle is cut off at every corner of such a square, the squares become octagons as faces of the new solid. Of the eight sides of each of these octagons, four are shared by other octagons, namely the ones lying on the edges of the original cube. This means that

12 edges of the solid are counted double if we simply add up the numbers of sides of the octagons. By this reasoning, we also obtain the number of edges of the solid as

$$6 \times 8 - 12 = 36.$$

Solution for Problem 9.10:

Of the five claims

(a) $A'C \perp BD$ (b) $AC \perp B'D'$ (c) $AD' \perp D'C$ (d) $AD' \perp B'C$
(e) $D'C \perp BC$

It is quite easy to see why claim (c) is incorrect. The three vertices A, D' and C of a cube are the corners of an equilateral triangle, as each of the line segments AD', $D'C$ and CA is a diagonal of a square face of the cube. These three segments are therefore of equal length, and therefore the sides of an equilateral triangle. Since the angles in an equilateral triangle are all equal to 60°, the lines are certainly not perpendicular. This is illustrated in the middle picture below, with the missing connection AC drawn as a dotted line.

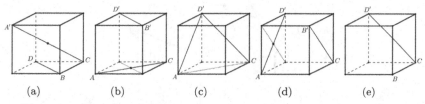

(a)   (b)   (c)   (d)   (e)

The other four pictures illustrate the other four claims, each of which is correct. This can be best understood by shifting one of the segments parallel to its original spot, through either an endpoint or the midpoint of the other.

Perhaps this is easiest to see for claim (B). Here, $B'D'$ has been shifted through the midpoint of AC, and this results in AC and this parallel being the two diagonals of the square $ABCD$, which are obviously perpendicular. Much the same is true for (D), where $B'C$ has been shifted through the midpoint of AD', again resulting in two diagonals of a square. For (E), such a shift is not required, since $D'C$ and BC already have the point C in common, and are obviously perpendicular.

The situation for claim (A) is a bit more subtle. In this case, it is probably easiest to consider BC shifted through the midpoint of $A'C$, which is simultaneously the midpoint if the cube. This results in the diagonal $A'C$

and the segment joining the midpoints of BB' and DD', and it is not too difficult to see that these are also perpendicular.

Solution for Problem 9.11:
First of all, let us consider the vertex K, as shown in the figure. Connecting K with any of the other points of the upper hexagon $GHIJKL$ certainly results in a line on a face. Similarly, connecting K with either D, E or F on the opposing hexagon also results in such a line. The only connections in the interior of the prism are those with A, B or C. We see that there are three line segments of the required type emanating from K. This is obviously also true of any of the other five corners of the upper hexagon, and since any segment of the required type must have an endpoint in one of these corners, the total number of connecting segments in the interior of the prism is therefore equal to

$$3 \times 6 = 18.$$

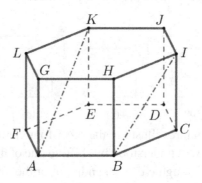

Solution for Problem 9.12:
If the object has a hexagonal outline, it cannot have been one of her round objects.

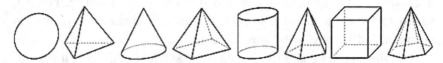

The sphere appears round, no matter which angle it is photographed from. The cone can only appear as a triangle if there is no round part to its outline, and similarly, the cylinder can only appear as a quadrilateral in such a case.

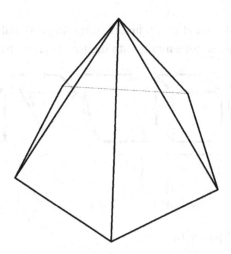

The other objects are all polyhedra with a certain number of vertices. If such a polyhedron has n vertices, its outline can never have more than n corners, no matter which angle it is viewed from. That eliminates the tetrahedron (the triangular pyramid), which only has four vertices, and the square pyramid, which only has five. We are left with the three objects on the right: the pentagonal pyramid, the cube and the hexagonal pyramid. The picture of the cube given here already has a hexagonal outline, so this is certainly possible for the cube. Viewing the hexagonal pyramid from the bottom means that we simply see its hexagonal base as its outline, so it is certainly possible to obtain a hexagonal outline in a photo of this object. This leaves us with the pentagonal pyramid. This does have six vertices, so it might be possible, but it might not be easy to see why such a pyramid can, indeed, have a hexagonal outline in a photo. Perhaps this picture will persuade you.

Solution for Problem 9.13:

The plan view (i.e. the view from above) of each of these objects is a square, so there is nothing to be gained there. Taking a close look at the pyramid (object D), however, we can note that there is only one direction in which we can view it as a square. The front view of this object is a triangle, and not a square.

This is a problem that really just requires some visualization. Here are the five given objects once more, with their respective front views below:

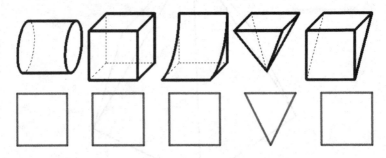

Solution for Problem 9.14:

Since the spider must climb up and cross over four rectangles en route from A to E, the shortest route can be found by flattening out the four rectangles to one big rectangle and then calculating the shortest distance as the length of the diagonal of the resulting big rectangle. In order to determine the length

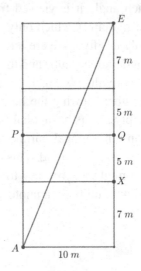

of the sides of this big rectangle, we need to first determine the lengths of the slanted roof edges from Q. This can be done by taking a close look at the side of the house.

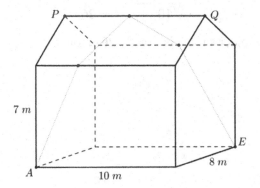

The slanted edge QX is the hypotenuse of the right triangle QXM. Since M is the midpoint of the horizontal line through X, we have $XM = 4$ m, and since the total height of the house is 10 m, we have $QM = 10 - 7 = 3$ m. This means that the length of QX is equal to

$$\sqrt{4^2 + 3^2} = 5.$$

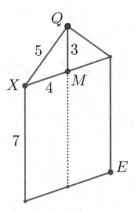

Since this is also the length of the slanted edge on the other side, we can now unfold the part of the house the spider must climb over, as shown in the figure. The distance from A to E is now seen to be the length of the diagonal of a rectangle with sides of length $10\,\text{m}$ and $7 + 5 + 5 + 7 = 24\,\text{m}$, and since

$$\sqrt{10^2 + 24^2} = 26,$$

we see that the spider's route is exactly $26\,\text{m}$ long.

Chapter 10

Money

We all know that money makes the world go round. It is therefore quite natural for a mathematics competition to contain problems dealing with money.

You will find that the statements in these problems involve different units of currency. One reason for this is the fact that Austria replaced its previously used currency 1 S (Austrian Schilling) = 100 g (Groschen) by the euro in 2002. In addition, you will find some fictional currencies showing up here and there.

PROBLEMS

Problem 10.1:

Two families, the Prozzis and the Schparzamms, are vacationing together. On their first evening, they go out for dinner. Mr. and Ms. Prozzi and their three children each have a luxurious surf and turf meal with lobster, steak and expensive red wine. The Schparzamm family is shocked. "We are each having an individual pizza and a cola. That comes to € 36 for the four of us. For the money you five are spending today, the four of us could have pizza and cola every day for three weeks!" Ignoring the health aspects of this for the moment, how expensive is the surf and turf luxury meal per person?

(S5K-13-7)

Problem 10.2:
Maria buys CDs at a price of 50 S for four and sells them at a price of 50 S for three. How many CDs does she have to sell to obtain a profit of 1,000 S?

(S5K-01-3)

Problem 10.3:
A photo lab charges a standard basic fee of 3 euros for any order. In addition, it charges 15 cents per picture for the first 10, and 9 cents per picture for all pictures from the eleventh on. Karl has 30 photos printed by the lab. How much does he have to pay per picture?

(StU-05-A2)

Problem 10.4:
Six children compare their weekly allowance. Alice and Boris together get 50 euros. The remaining four children get an average of 7 euros. What is the average allowance of all six children?

(S5K-13-9)

Problem 10.5:
The post office of Oberschnatterbach only has stamps in denominations of 6 Schillings and 9 Schillings. A package needs to be posted with stamps in a total value of 120 Schillings. In how many different ways is it possible to do this?

(O-97-1)

Problem 10.6:
Pam and Peggy are negotiating with their parents about their allowances. Pam agrees with her parents to receive 10 euros in the first month, 15 euros in the second and 5 euros more in each following month than in the month preceding it. Peggy agrees with her parents that she will receive 2 euros in each of the first two months, then 4 euros in each of the next two months, then 8 euros for two months, and so on. (In other words, her pocket money will be doubled every two months.) How much will each girl get in the 12th month? Who will get more money over the course of the whole year and how much more will she get?

(StU-09-B1)

Problem 10.7:
According to the somewhat antiquated laws of Erbenstein, the assets of a father must be divided after his death in such a way that his widow receives half as much as a son, but twice as much as a daughter. A wealthy man leaves 9,009 talers to his wife, his two sons and his daughter. How much must each of them receive?

(StU-01-B6)

Problem 10.8:
The Huber family — the mother, the father and their children — is on a trip to the museum. Both the admission fee for children and the admission fee for adults in euros is an integer amount greater than 1 euro. The fee for an adult is three times the fee for a child. In total, the family pays 27 euros for admission. How many children are there in the Huber family?

(StU-11-B4)

Problem 10.9:
Grandpa has won 25,200 euros in the lottery. He plans to distribute his winnings in the following way. Each of his five grandchildren will get the same amount, and each of his two sons will get twice as much as each of the grandchildren. Finally, Grandpa wants to keep enough of the money for a holiday in the Caribbean with Grandma, namely one-third of the total winnings, reduced by the amount one son will receive. How much money will remain for the Caribbean holiday?

(StU-06-B2)

Problem 10.10:
Sandra collects coins in a can. She has 10 cent, 20 cent and 50 cent coins worth a total of 9 euros. She has an equal number of 10 cent coins and 20 cent coins, and three times as many 50 cent coins as she has 10 cent coins. How many 20 cent coins does she have in the can?

(StU-15-B7)

Problem 10.11:
Several coins are lying on a table, valued at 2 euros, 1 euro or 50 cents each. The total value of the coins is € 25. Pierre notices that there are 20 coins

on the table altogether, and that there are exactly 2 more € 2 coins than € 1 coins. How many 50 cent coins are there on the table?

(S5K-13-7)

Problem 10.12:
For a school play, the auditorium is equipped with 300 seats. Tickets to the show cost 50 S for adults and 25 S for students. Unfortunately, the show was not sold out, but the income from the ticket sales came to 12,000 S. Georg claims that exactly 160 of the seats were sold to adults. Explain why he cannot be correct in saying this. What is the smallest number of adults who must have bought tickets?

(StU-01-B4)

Problem 10.13:
Four apples cost 3 S less than five bananas, whereas seven apples cost 1.50 S more than five bananas. How much does one banana cost?

(StU-00-A2)

Problem 10.14:
Apple Joe sells his apple juice in bottles. Some of them contain 0.7 liters and some contain 1 liter. He plans to sell every 0.7 liter bottle for 10 Schillings and isn't sure how expensive a 1 liter bottle should be, but the price (in Schillings) should be an integer. He ends up selling exactly 33 liters of juice for exactly 456 Schillings. How many bottles of each size did he sell and how expensive was each 1 liter bottle?

(O-01-4)

Problem 10.15:
Mato is out shopping. Comic books are € 2 each, laser pointers are € 3 each and CDs are € 7 each. He buys at least one of each of them. In total, he buys seven items for a total of € 31. How many laser pointers does he buy?

(StU-09-A5)

Problem 10.16:
At the snack bar in my school, a donut costs 70 cents, a sandwich costs 90 cents, and a slice of pizza costs 1.30 euros. Each member of a group of students buys exactly one snack. After that, the salesperson realizes that she

has sold at least one piece of each of the three kinds of snack, and that she has taken in a total of 6 euros. How many students bought a snack?

(StU-06-A8)

Problem 10.17:
If I buy 10 packs of gummi bears at my local store, I pay less than 12 euros. If I buy 11 packs, I pay more than 13 euros. How much does one pack of gummi bears cost at my local store, assuming that there is no special deal for large purchases?

(O-03-3)

Problem 10.18:
Anne and Franz are buying stamps. Anne buys 7 Schilling stamps and 12 Schilling stamps, and she pays a total of 116 Schillings for 13 stamps. How many stamps of each kind does she buy? Franz buys 16 stamps, valued at 9 Schillings or 13 Schillings each. He pays with two 100 Schilling bills and receives a 20 Schilling bill and several 5 Schilling coins as change. How many 13 Schilling stamps does he buy? How many 5 Schilling coins does he receive as change?

(StU-99-B3)

Problem 10.19:
Stimpy has an unlimited supply of stamps valued at 51 cents and 68 cents each. He would like to send a package, which costs 5.78 euros postage. He needs to know how many stamps of each type he will need. Determine all possible solutions of this problem.

(O-02-1)

Problem 10.20:
Hans has a 500 Schilling bill. He asks Franz to give him change in bills of smaller denominations. Franz says "I have more than 500 Schillings, and I only have bills valued at 20, 50 and 100 Schillings, but unfortunately, I am not able to give you change for your 500 Schilling bill". How much money can Franz have at most?

(O-97-4)

TIPS

Tip for Problem 10.1:
Determine the total amount the Schparzamms would have to pay every day for a daily pizza and cola for each family member over the course of three weeks.

Tip for Problem 10.2:
Calculate Maria's profit when she sells 1 CD.

Tip for Problem 10.3:
Calculate the total costs for Karl's order first.

Tip for Problem 10.4:
Remember that the average is calculated by dividing the sum of the values by their number.

Tip for Problem 10.5:
Obviously, it is possible to post the package with twenty 6-Schilling stamps. Other options include replacing a specific number of 6-Schilling stamps with 9-Schilling stamps.

Tip for Problem 10.6:
The first part of this problem is solvable by straightforward calculation. The calculation of the sums they get in the first year becomes easier if their monthly allowances are grouped in a smart way.

Tip for Problem 10.7:
Since the widow receives twice as much as a daughter, and a son receives twice as much as the widow, a son receives four times as much as a daughter.

Tip for Problem 10.8:
Since the ticket price for an adult is three times the admission for a child, it makes sense to treat father and mother as three children each.

Tip for Problem 10.9:
A common way to solve such problems is to set up an equation. Suppose that each grandchild will receive x euros. Then express the shares of each of the family members in terms of x and solve the resulting equation. If you prefer, it is also possible to determine the share of each grandchild by direct calculation.

Tip for Problem 10.10:
Again, the problem can be solved by an equation.

Tip for Problem 10.11:
See the previous problem!

Tip for Problem 10.12:
Even if all 300 seats had been sold to students, ticket revenue would not have been more than 7,500 S. Therefore, for a total income of 12,000 S, a certain number of adult tickets must have been sold. Try to find arguments why this number must exceed 160.

Tip for Problem 10.13:
First, try to determine the cost of an apple from the information given. Alternately, a standard method of solving this type of problem is to set up a system of equations.

Tip for Problem 10.14:
Note that the amount of apple juice sold by Joe (in liters) is an integer. This restricts the possible numbers of 0.7 liter bottles he might have sold.

Tip for Problem 10.15:
Buying exactly one of each of the items, Mato spends € 12 for 3 items, so we know how much he spends for four more items. The rest can be done by some trial and error involving inequalities.

Tip for Problem 10.16:
This problem is rather similar to the previous one, and can be solved in a very similar way.

Tip for Problem 10.17:
Note that the price in cents of a pack of gummi bears must be integer. Based on the given information, two inequalities can be formulated to calculate this price.

Tip for Problem 10.18:
A standard way to solve the first part is to establish and solve a system of equations: For this purpose, let n be the number of 7 S stamps and let m be the number of 12 S stamps that Anne buys. There are other ways to do this as well, if you prefer. For the second part, express the amount Franz has to

pay in terms of the numbers of 9 S stamps and 13 S stamps he buys. This value must be divisible by 5 because of the given information.

Tip for Problem 10.19:
Establish an equation in x and y, where x is the number of 51 cents stamps and y is the number of 68 cents stamps.

Tip for Problem 10.20:
If Franz had a sufficiently large number of 20 Schilling bills and 50 Schilling bills, he could make change for almost any amount.

SOLUTIONS

Solution for Problem 10.1:
Since one day of pizza and cola for each of the Schparzamms costs € 36 and three weeks are 21 days, they would have to pay $21 \times$ € $36 =$ € 756. If this is the amount the five members of the Prozzi family spend for their luxury meal, the expenses per person are € $756 \div 5 =$ € 151.20.

Solutions for Problem 10.2:

Solution 1:
Maria's profit when she sells one CD (in S) is equal to $\frac{50}{3} - \frac{50}{4} = \frac{50}{12}$. Therefore, the number of CDs she has to sell for a profit of 1,000 S is equal to $1,000 \div \frac{50}{12} = 240$.

Solution 2:
Maria buys CDs at a price of 50 S for four and sells them at a price of 50 S for three. Since the least common multiple of 3 and 4 is 12, it makes sense to calculate prices when she buys or sells 12 CDs, respectively. For each 12 CDs she buys, she has to pay $3 \times 50 = 150$ S, and for any 12 CDs she sells, she takes in $4 \times 50 = 200$ S. Thus, trading 12 CDs, she makes a profit of 50 S. Consequently, because, $1000 \div 50 = 20$, she must sell $20 \times 12 = 240$ CDs for a profit of 1,000 S.

Solution for Problem 10.3:
The total costs for Karl's order contain 3 euros basic fee, 10×0.15 euros $= 1.50$ euros for the first 10 pictures and 20×0.09 euros $= 1.80$ euros for the remaining 20 pictures. Therefore, he pays 6.30 euros for his order, and he therefore pays 6.30 euros $\div 30 = 0.21$ euros $= 21$ cents per picture.

Solution for Problem 10.4:

Altogether, the children get 50 euros $+ 4 \times 7$ euros $= 78$ euros per week. It follows that the average pocket money of all six children is equal to 78 euros $\div 6 = 13$ euros.

Solution for Problem 10.5:

The package can be mailed using twenty 6-Schilling stamps. Since $18 = 3 \times 6 = 2 \times 9$, any three 6-Schilling stamps can be replaced by two 9-Schilling stamps. Thus, 3, 6, 9, 12, 15 or 18 6-Schilling stamps can be replaced by 2, 4, 6, 8, 10 or 12 9-Schilling stamps, offering six more options. Therefore, there are seven ways to send the package with stamps totaling 120 Schillings.

Solution for Problem 10.6:

Up until the 12th month, Pam's allowance is raised by 5 euros eleven times, and so the amount she gets in the 12th month is equal to €10 $+ 11 \times$ €5 $=$ €65. In the same period, Peggy's allowance is doubled five times, and so she gets $2^5 \times$ €2 $=$ €64 in the 12th month.

Since Pam's allowance is raised by 5 euros eleven times, we can write the total amount of money she gets as

$$\text{€}10 + \text{€}15 + \text{€}20 + \cdots + \text{€}60 + \text{€}65.$$

The sum of the amounts that Pam gets in the first and twelfth month, in the second and the eleventh month, and so on always sum to

$$\text{€}10 + \text{€}65 = \text{€}15 + \text{€}70 = \cdots = \text{€}75.$$

She gets this amount six times, and so the total amount she gets in the whole year is $6 \times$ €75 $=$ €450. On the other hand, Peggy gets a total of

$$\text{€}2 \times (2 + 4 + 8 + 16 + 32 + 64) = \text{€}252.$$

We see that Pam will get €450 $-$ €252 $=$ €198 more than Peggy in the course of the year.

Solution for Problem 10.7:

We don't want to dwell on the unfair distribution here, of course. We can hope that modern societies would find better ways of dealing with such things than they did in old Erbenstein.

The widow receives twice as much as the daughter, and the two sons together receive eight times as much as the daughter. The daughter therefore receives one-eleventh of the assets of the father. We see that the daughter's share is $9,009 \div 11 = 819$ talers, since the mother's share is twice that, she receives 1,638 talers, and the share of each of the two sons is twice that again, and they each receive 3,276 talers.

Solution for Problem 10.8:
Since the admission for an adult is three times the fee for a child, the parents together pay the same admission fee as six children would. Consequently, the Huber family, with its n children, pays the same admission fee as a group of $n + 6$ children would. The admission fee for a child is therefore equal to $27 \div (n + 6)$, which must be an integer greater than 1. The only divisor of 27 fulfilling this requirement is 9, and so we have $n = 3$. The number of children in the Huber family is therefore three.

Solutions for Problem 10.9:

Solution 1:
Suppose each of the five grandchildren receives x euros. It follows that each of the two sons will receive $2x$ euros, and the amount that Grandpa will keep is equal to $25,200 \div 3 - 2x = 8,400 - 2x$ euros. We therefore obtain the equation

$$5x + 2 \times 2x + 8,400 - 2x = 25,200,$$

which reduces to $7x = 16,800$ or $x = 2,400$. This means that Grandpa will have € $8,400 -$ € $4,800 =$ € $3,600$ left over.

Solution 2:
One-third of the total winnings is equal to € 8,400. Since this amount is sufficient for the holiday in the Caribbean and for the share of one of the two sons, the remaining € 16,800 of the winnings is shared among the five grandchildren and a son who gets twice as much as a grandchild. This means that each grandchild gets one-seventh of € 16,800, or € 2,400. Each son gets € 4,800, and therefore € $8,400 -$ € $4,800 =$ € $3,600$ remains for the Caribbean holiday.

Solution for Problem 10.10:
Let x be the number of 10 cent coins. Then the number of 20 cent coins is x as well, and the number of 50 cent coins is $3x$. The total value of Sandra's coin collection is then

$$10 \times x + 20 \times x + 50 \times 3x = 180 \times x.$$

Since 9 euros are equal to 900 cents, we have the equation $180x = 900$, or $x = 5$. It follows that Sandra has five 20 cent coins in the can.

Solution for Problem 10.11:
Let x be the number of 1 euro coins. Then the number of 2 euro coins is equal to $x + 2$. The total number of coins on the table is 20, and so the number of 50 cent coins is equal to

$$20 - x - (x + 2) = 18 - 2x.$$

Considering the total value of all coins, we obtain the equation

$$x + 2 \times (x + 2) + 0.5 \times (18 - 2x) = 25,$$

which reduces to $2x + 13 = 25$, or $x = 6$. The number of 50 cent coins is therefore $18 - 2 \times 6 = 6$.

Solution for Problem 10.12:
Suppose George is right. Then the income from adult ticket sales is 8000 S. The remaining amount of 4000 S must therefore come from the sale of tickets to students. The number of students in the auditorium is therefore equal to $4{,}000 \div 25 = 160$. This results in a total of $160 + 160 = 320$ visitors to the show, which is a contradiction to the maximum capacity of 300 of the auditorium.

Let x be the number of tickets sold to adults, and let y be the number of tickets sold to students. Then the total ticket revenue is equal to $50x + 25y = 12{,}000$, and we therefore have $2x + y = 480$. Since the show was not sold out, we have $x + y < 300$, and therefore $x > 180$. At least 181 tickets must have been sold to adults.

Solutions for Problem 10.13:

Solution 1:
Four apples cost 3 S less than five bananas, and seven apples cost 1.50 S more than five bananas. It follows that $7 - 4 = 3$ apples cost 4.50 S,

and consequently one apple costs 1.50 S. This means that four apples cost $4 \times 1.50 = 6$ S. Five bananas therefore cost $6 + 3 = 9$ S, and we see that one banana costs $9 \div 5 = 1.8$ S.

Solution 2:

Let x and y be the costs of one banana and one apple (in S), respectively. Since four apples cost 3 S less than five bananas, we obtain the equation

$$4y = 5x - 3$$

and since seven apples cost 1.50 S more than five bananas, we obtain the equation

$$7y = 5x + 1.5.$$

Dividing the first equation by 4 and the second by 7, we therefore obtain

$$y = \frac{5x - 3}{4} = \frac{5x + 1.5}{7}.$$

Solving this equation for x, we obtain $35x - 21 = 20x + 6$, which simplifies to $15x = 27$ or $x = \frac{27}{15} = \frac{9}{5} = 1.8$. A banana therefore costs 1.8 S.

Solution for Problem 10.14:

Since the amount of apple juice sold by Joe (in liters) is an integer, the number of 0.7-liter bottles he sold must be a multiple of 10 less than $33 \div 0, 7$, and thus less than 47. This means that the total amount of apple juice sold in 0.7-liter bottles is a multiple of 7. He therefore sold either 10, 20, 30 or 40 small bottles containing a total of 7, 14, 21 or 28 liters of juice for a total price of 100, 200, 300 or 400 Schillings. In any case, the remaining amount of 26, 19, 12 or 5 liters of juice was sold in 1-liter bottles. Therefore, the remaining revenue of 356, 256, 156 or 56 Schillings, divided by the number of 1-liter bottles, gives the price for one 1-liter bottle. Since no division besides $156 \div 12 = 13$ yields an integer result, Joe sold thirty 0.7-liter bottles and twelve 1-liter bottles, and the price for a 1-liter bottle was 13 Schillings.

Solution for Problem 10.15:

Buying exactly one of each of the items, Mato spends € 12 for 3 items. He therefore spends € 19 for 4 more items. Obviously, € 19 is not enough to buy 3 CDs. On the other hand, if he only buys one CD, he would spend € 10 for three more items, which contradicts the fact that the price for any of these items is at most € 3. He therefore spends € 14 for two more CDs and € 5 for the remaining two items. This means that he buys one more laser pointer, for a total of two.

Solution for Problem 10.16:
Since she has sold at least one piece of each of the three kinds of snack, she obviously took in 2.90 euros for one donut, one sandwich and one slice of pizza from three students. Since any kind of snack costs an odd multiple of 10 cents, and she took in a total of 3.10 euros from the rest of the group, which is also an odd multiple of 10 cents, the number n of the remaining students who made a purchase is odd. Obviously, n is larger than 1. On the other hand, since each student bought exactly one snack, n cannot be larger than 3, since 5 pieces of the least expensive snack cost 3.50 euros. It follows that n is equal to 3. From these three students, she took in a total of 3.10 euros, i.e. 20 cents more than the price of one donut, one sandwich and one slice of pizza. This extra cost of 20 cents can only be caused by buying a sandwich instead of a donut. Consequently, she took in 3.10 euros for two sandwiches and one slice of pizza. Altogether, we see that six students bought a snack; 1 donut, 3 sandwiches and 2 slices of pizza for 6 euros.

Solution for Problem 10.17:
Let p be the price of one pack of gummi bears in cents. From $10\,p < 1200$, we get $p < 120$. On the other hand, from $11p > 1300$, we get $p > 1300 \div 11 > 118$. Since p is an integer, we therefore have $p = 119$. The price of a pack of gummi bears is therefore 1.19 euros.

Solution for Problem 10.18:
For the first part, we notice that Anne would have to pay $13 \times 7\,S = 91\,S$ for thirteen 7 S stamps. Each 12 S stamp she buys instead of a 7 S stamp raises the price by 5 Schillings. In total, the amount she must pay is raised by $116\,S - 91\,S = 25\,S$, and she therefore buys five 12 S stamps and thus eight 7 S stamps.

Suppose, Franz buys n 13 S stamps and, since he buys 16 stamps in total, $(16 - n)$ stamps valued at 9 S. The price to be paid for these sixteen stamps in Schillings is

$$13\,n + 9(16 - n) = 4\,n + 144,$$

which is obviously an integer not less than 144, which is divisible by 4. Since Franz receives one 20 S bill and some 5 S coins as change, this number must be a multiple of 5 less than 180 as well. The only number fulfilling these requirements is 160, which gives us $4n + 144 = 160$ and therefore $n = 4$. Franz therefore buys four 13 S stamps and twelve 9 S stamps for a total of 160 S. He gets 20 S change in the form of 5 S coins, and so he receives four 5 Schilling coins as change.

Solution for Problem 10.19:
The postage required to mail the package is equal to 578 cents. Let x denote the required number of 51 cent stamps, and let y denote the required number of 68 cent stamps. For the total cost, we get the equation $51x + 68y = 578$. This equation can be divided by 17, yielding $3x + 4y = 34$ or

$$y = \frac{34 - 3x}{4} = \frac{36 - (3x + 2)}{4} = 9 - \frac{3x + 2}{4}.$$

Both x and y must be non-negative integers, and therefore $\frac{3x+2}{4}$ must be a positive integer not greater than 9, and consequently $x \in \{2, 6, 10\}$. The corresponding values for y are 7, 4 and 1, respectively. This means that Stimpy might use either two 51 cent stamps and seven 68 cent stamps, or six 51 cent stamps and four 68 cent stamps, or ten 51 cent stamps and one 68 cent stamp.

Solution for Problem 10.20:
From the information given, it is clear that 400 Schillings is the largest amount divisible by 100 that Franz can pay using his bank notes. It does not matter whether he pays any amount of 100 Schillings using five 20 Schilling bills, two 50 Schilling bills or one 100 Schilling bill. In addition, he can have at most four 20 Schilling bills and at most one 50 Schilling bill, so Franz can have at most 530 Schillings.

Chapter 11

It's About Time

The passing of time is a surprisingly multi-faceted thing. There are interesting questions to be asked about brief snippets of time or long stretches, as well as just the passing of an indeterminate amount of time. To reflect this, this section is sub-divided into several bits, each of which contains problems of a certain type.

PROBLEMS

Clocks and Clock Faces

Problem 11.1:
The minute hand has fallen off the clock in the figure. What time does the clock show?

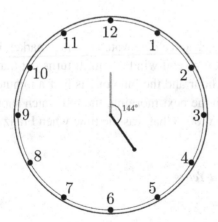

(S5K-01-4)

Problem 11.2:
How large is the angle determined by the hands of a clock at 12:15?

(S5K-03-10)

Problem 11.3:
How large is the angle determined by the hands of a clock at 11:48?

(StU-97-A5)

Problem 11.4:
The display of a digital clock has four digits and shows the time from 00:00 to 23:59 every day. Determine the fraction of the time each day in which the digit 2 is visible somewhere on the clock.

(O-98-6)

Problem 11.5:
If a digital clock shows the time as 8:02, it is possible to read this as the division $8 \div 2$ and calculate the result as being equal to 4 without leaving a remainder. In this way, the time 9:11 yields 0 with remainder 9, and 18:12 yields 1 with remainder 6. How often between 8:30 and 11:30 does the division leave the remainder 0?

(S5K-03-7)

Problem 11.6:
Franz buys a red watch and a blue watch at a flea market. He sets them both to the same correct time and winds them. It turns out that the red watch is a minute fast each hour and the blue one is half a minute slow each hour. When Franz got up the next morning, the red watch showed 6:30 and the blue watch showed 6:00. What was the time when Franz set the watches?

(StU-96-A4)

Weeks, Months and Years

Problem 11.7:
1 stands for January, 2 stands for February, 3 stands for March, and so on. What is the sum of all numbers that stand for the months in a year?

(StU-17-A1)

Problem 11.8:
In a regular year (that is to say, a non-leap-year), January 13th falls on a Friday. Which month will have the next "Friday, the 13th"?

(StU-04-A5)

Problem 11.9:
There are 13 Fridays in the months May, June and July in a certain year. Which days of the week can August 13th of that year be on?

(StU-01-A4)

Problem 11.10:
In a certain year, the months July, August and September have a total of 14 Sundays. Which day of the week is March 28th of that year?

(StU-99-A3)

Year Numbers

Problem 11.11:
The year 1991 was a palindromic year, because its number reads the same forwards and backwards. How many years pass between the next two palindromic years after this?

(O-95-1)

Problem 11.12:
The number of the year 1998 has the property that it is twice a number (999) composed of digits that are all the same. When was this last the case before 1998, and when will it be the case the next time after 1998?

(O-98-1)

How Old Am I?

Problem 11.13:
Karl, his mother and his sister Anna are 56 years old in total. When Karl was as old as his sister is now, his mother was 32 years old. How old is Karl now?

(O-03-4)

Problem 11.14:
Christian's age in the year 2001 is equal to the sum of the digits in the year he was born. How old is Christian?

(S5K-01-15)

Problem 11.15:
Grandfather Joseph, his son Gerhard and grandson René are asked their ages. The grandfather answers: "The sum of Gerhard's age and mine is 109 years and the sum of Gerhard's and René's is 56 years. Together, René and I are 85 years old". How old is his son Gerhard?

(S5K-02-6)

Problem 11.16:
Mr. Dreher's age together with that of his wife will be equal to 108 when his wife is as old as he is now. How old is Mr. Dreher and how old is his wife, if we know that Mr. Dreher is now twice as old as his wife was when he was the age his wife is now?

(StU-92-9)

Problem 11.17:
Mr. Hofer says "If I multiply the ages of all my children (in integer numbers of years), I get 660!". None of my children is older than 20 and one is 4 years younger than another one. If I multiply the age of my youngest with the age of my oldest, I get my mother's age. How old can his children be according to this information? Determine all possible solutions.

(StU-99-B5)

Problem 11.18:
Which of the following numbers is the closest to your age in seconds?
5,000,000,000, 500,000,000, 50,000,000, 5,000,000, 500,000

(StU-98-A7)

Problem 11.19:
Tony Milliard just turned 1,000,000,000 (1 milliard, or 1 billion, depending on where you live) seconds old. His age in years is

(a) less than 5   (b) between 5 and 20   (c) between 20 and 35

(d) between 35 and 50   (e) more than 50

(StU-08-A2)

As Time Goes By

Problem 11.20:

The average time of the three medallists at a 100 m sprint race is 10.5 seconds. The average time of the other participants is 13.2 seconds and the average time of all participants is 12.39 seconds. How many participants were there?

(O-99-5)

Problem 11.21:

Christoph is riding his bike along the tram tracks. He notes that a tram comes from the opposite direction every 18 minutes and that he is passed by one going in the same direction as he is every 22 minutes. How many minutes pass between departures of trams going in either direction, assuming constant speeds both for Christoph and for all the trams?

(S5K-00-1)

Problem 11.22:

Having just shaken his pursuers, Joe Welles, super-agent 008, runs into the Borovitzkaya Station, where he sees a train leaving the platform just as he enters. Joe has lost his watch during the pursuit. A glance at the timetable shows him that a train to Krilatskoye leaves every 12 minutes all day long, starting from 0:06. Another train to Praschskaya leaves every 15 minutes from 0:03 and one to Jugo-Sapadnaya leaves every 10 minutes from 0:02. What is the longest interval he might have to wait until the next train leaves the platform, assuming they are all running on schedule?

(S5K-08-7)

Problem 11.23:

A traveler asks a man what the time is. The man answers: "Until the end of the day, there still remain three times two-ninths of the time that has already passed from the beginning of the day". Will the traveler be able to reach his train, if he knows that it leaves at 14:45, and it will take him 12 minutes to reach the platform?

(StU-92-7)

Problem 11.24:

Eugene's parents will buy him a gaming console if he helps out with the housework for an average of 6 hours a week over the next 7 weeks. In the first five weeks, he helps out for 8 hours, 3 hours, 6 hours, 7 hours and 7 hours again, respectively. How many hours on average will he have to help out during the next two weeks in order earn his prize?

(S5K-09-8)

Problem 11.25:
A plane leaves for Chicago every 40 minutes at my local airport. Every 60 minutes, a plane leaves for New York and every 72 minutes a plane leaves for Los Angeles. At 12:00, there are planes leaving in all three directions. At what time was this last the case?

(StU-02-A1)

Problem 11.26:
Noah lives in Kumberg in Austria, and would like to phone his uncle Barack in Washington after school. When Noah gets up at 6 in the morning, it is midnight in Washington. Noah has classes until 13:20, and can't phone before that. He goes to bed at 20:00. Uncle Barack starts work in Washington at 8:30, and can't be reached by phone until 21:00 in the evening. How many minutes long is the daily window during which Noah can phone his uncle?

(S5K-16-8)

Problem 11.27:
Five patients arrive at a doctor's office simultaneously. He knows that their treatments generally take 5 minutes, 7 minutes, 10 minutes, 12 minutes and 15 minutes, respectively. The doctor sees them in an order that guarantees that their total waiting times are as small as possible. How long is their total waiting time?

(StU-96-A5)

Problem 11.28:
Niklas and Paula run several laps around a lake. They start together at the hollow tree and run until they arrive simultaneously at that spot again for the first time. Paula runs a lap around the lake in 12 minutes, but it only takes Niklas 10 minutes. How many minutes will they run until they stop?

(S5K-17-1)

Problem 11.29:
A traffic light shows red for 40 seconds, then red-yellow for 3 seconds, and then green for 45 seconds. It then blinks green for 4 seconds and then shows yellow for 3 seconds before switching back to red for the next 40 seconds, restarting the cycle. At exactly 7 a.m., the light changes from yellow to red. What is showing at exactly 7:30 a.m.?

(StU-03-B1)

Problem 11.30:
The traffic light at an intersection shows green for 50 seconds, then yellow for 5 seconds and then red for 30 seconds. It then immediately changes back to green. The light changes to green at exactly 7 a.m. How often does the light change to green between 7 a.m. and 7 p.m.?

(StU-17-B3)

TIPS

Tip for Problem 11.1:
The hour-hand has rotated 144° from the vertical position it held at 12:00. Think about the angle it rotates each full hour. From one full hour to the next, the minute-hand rotates all the way around, back to the vertical position, and in any given fraction of a full hour, it rotates that same fraction of the full circle from the vertical position.

Tip for Problem 11.2:
Note that the hands of the clock coincide at 12:00. In the 15 minutes from then to 12:15, the minute-hand moves from a vertical position to a horizontal one. How much does the hour-hand move in that time?

Tip for Problem 11.3:
Note that this problem is almost the same as the last one. The difference is that the time is not 15 minutes after the hands of the clock coincide, but rather 12 minutes before they do.

Tip for Problem 11.4:
In order to solve this, we will have to consider the times when the digit 2 is visible as part of the hour-digits separately from the times when it is not.

Of course, there are also times when the digit 2 is displayed as one of the minute-digits, but not one of the hour-digits.

Tip for Problem 11.5:
The hours in this problem are 8, 9, 10 and 11. In each case, we want to count the number of divisors of the number. But, watch out! Not all divisors of all four of these numbers correspond to times in the interval we are asked to consider.

Tip for Problem 11.6:
We know how much the times displayed on each of the watches drifts away from the correct time each hour. It is therefore not difficult to figure out how much the displayed times *n* the two watches drift apart from each other each hour. Since we know how far apart the displayed times are in the morning, it is not difficult to figure out how much time has elapsed since the watches were set.

Tip for Problem 11.7:
Of course, you can just add up all the numbers from 1 to 12 to get the answer. But take a look at the sum $1 + 2 + 3 + \cdots + 10 + 11 + 12$. Note that the sum of the first and last numbers in this expression is 13. This is also the sum of the second and next to last numbers. Noting this, can you say what the sum of all 12 numbers must be without actually doing the addition?

Tip for Problem 11.8:
In order to solve this, you may have to look at a calendar. On the other hand, you could recall the old saying "Thirty days hath September, April, June, and November; all the rest have thirty-one, except for February alone". This should help count through the months. Of course, the problem becomes even easier if you know a little modular arithmetic.

Tip for Problem 11.9:
First of all, you will need to check out how many days there are altogether in the months May, June and July. On which date could the first of the 13 Fridays possibly fall?

Tip for Problem 11.10:
This problem is quite similar to the last one, but it may seem to be impossible at first glance. It isn't, though. Just add up the days in July, August and September.

Tip for Problem 11.11:
The first two digits of 1991 are 19. In any palindromic year, the first two digits determine the last two. This means that you already know what the next two palindromic years will be!

Tip for Problem 11.12:
The last year before 1998 with this property was twice a three-digit number, while the next larger one cannot be, as 999 is the largest three-digit number with identical digits. The next larger one must therefore be twice a four-digit number with this property.

Tip for Problem 11.13:
This is a problem where it will be quite useful to introduce variables for things you don't know. Try setting Karl's age as k and Anna's age as a. What do you know about their mother's age?

Tip for Problem 11.14:
What decade might he have been born in? You can get this by trying out the different cases.

Tip for Problem 11.15:
Try adding up all three of the sums we are given. What does this number represent?

Tip for Problem 11.16:
This one isn't really hard, just confusing. You want to know Mr. Dreher's age, so call it d, and his wife's age, so call that w. Now read through the text again. The first sentence will give you an equation, and the second half of the last one will give you another.

Tip for Problem 11.17:
It will be helpful to consider the prime decomposition of 660 here. Note that none of his children can be one year old, since the product of the ages of the youngest and oldest child could then not be the age of his mother.

Tip for Problem 11.18:
Note that a minute has 60 seconds, an hour has 60 minutes, a day has 24 hours and a year has 365 days (or thereabouts...). Now all you need to know is how old you are as a potential contestant.

Tip for Problem 11.19:
If you figured out how many seconds there are in a year for the last problem, you are already well on your way to solving this one as well.

Tip for Problem 11.20:
Since we want to know the number of participants, this number is a good candidate for a variable x.

Tip for Problem 11.21:
This is a really tricky one, so watch out! You may be inclined to think that the answer might just be the average of the two given numbers. That would make the answer $\frac{22+18}{2} = 20$ minutes, right? Unfortunately, that isn't the correct answer! Try introducing variables for things you don't know, but would like to. One good option is to introduce the variable t as the speed of the tram, b as the speed of the bike and x as the length of the intervals we would like to determine.

Tip for Problem 11.22:
Note that there are 60 minutes in an hour, and 12, 15 and 10 are all divisors of 60. This means that we just have to consider the minutes of the trains' various departure times, ignoring the hours.

Tip for Problem 11.23:
A day has 24 hours, of course. You don't know how much time has passed since midnight, so just assume that it has been x hours since then. That should get you there.

Tip for Problem 11.24:
If Eugene has to help for 6 hours a week on average over 7 weeks, how many hours will he have to help out altogether?

Tip for Problem 11.25:
This one should become easy to solve if you think of it as a least common multiple problem.

Tip for Problem 11.26:
What time is it in Kumberg when Uncle Barack gets off work? What time is it in Kumberg when he starts work again?

Tip for Problem 11.27:
If the doctor sees the patient first who will take the longest, everyone else will have to wait the longest, so that doesn't seem like a good idea. The opposite should do the trick, though.

Tip for Problem 11.28:
Comparing this problem to 11.25 should give you a hint.

Tip for Problem 11.29:
First of all, calculate how long a full cycle of the traffic light is, from the time it changes to red until it changes to red again. After that, the problem reduces to simple division.

Tip for Problem 11.30:
This last one of the section is very similar to the previous one. Once you know the length of the full cycle, simple division and some common sense will give you the answer.

SOLUTIONS

Solution for Problem 11.1:

We know that the hour-hand of a clock rotates all the way around, or by a full angle of 360°, in 12 hours. This means that it rotates $360° \div 12 = 30°$ in an hour. In our problem, the hour-hand has rotated by 144° since 12:00, and this means that the number of hours that have passed since 12:00 is equal to $144 \div 30 = 4\frac{24}{30}$, or $4\frac{4}{5}$. The minute-hand has therefore rotated four-fifths of the full angle from the vertical position since it was last there, and since $\frac{4}{5} \times 360° = 288°$, the angle of the minute-hand to the vertical is equal to 288° in a clockwise direction (which is, of course, equal to $360° - 288° = 72°$ in a counter-clockwise direction). The number of minutes that have passed since 4:00 is therefore equal to $\frac{4}{5} \times 60 = 48$, and the time showing on the clock is therefore 4:48.

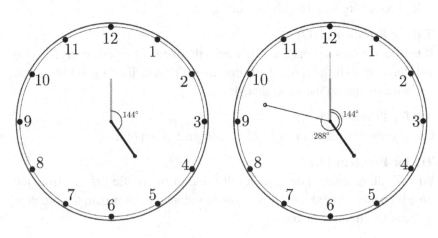

Solution for Problem 11.2:

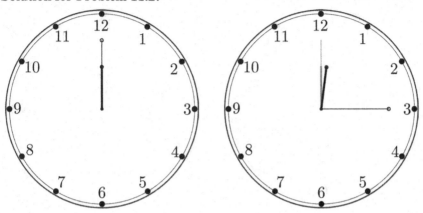

Let us take a close look at the figures above. In the left-hand figure, we see a clock showing 12:00, and in the right-hand figure, we see one showing 12:15. One-quarter of an hour has passed from the first situation to the second, and in this time, the minute-hand has moved one-quarter of the full angle of 360° it rotates each full hour, or $\frac{1}{4} \times 360° = 90°$. During that same time, the hour-hand has rotated by an angle equal to one-quarter of the angle it rotates each hour. The angle it rotates in a full hour is equal to 30°, since it rotates by a full angle of 360° in 12 hours, and thus $360° \div 12 = 30°$ in an hour. The angle it rotates in one-quarter of that time is therefore equal to $\frac{1}{4} \times 30° = 7.5°$. We therefore see that the angle contained by the hour-hand and the minute-hand at 12:15 is equal to $90° - 7.5° = 82.5°$.

Solution for Problem 11.3:

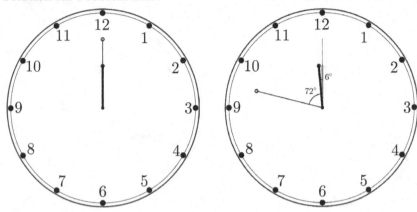

As in the last problem, we can take a close look at the figures. In the left-hand figure, we once again see a clock showing 12:00. In the right-hand figure, we now one showing 11:48. This is 12 minutes before 12:00, or $\frac{12}{60} = \frac{1}{5}$ of an hour. In this time, the minute-hand moves one-fifth of the full angle of $360°$ it rotates each full hour, or $\frac{1}{5} \times 360° = 72°$. During that same time, the hour-hand rotates by an angle equal to one-fifth of the $30°$ it rotates each hour, or $\frac{1}{5} \times 30° = 6°$. We therefore see that the angle contained by the hour-hand and the minute-hand at 11:48 is equal to $72° - 6° = 66°$.

Solution for Problem 11.4:

First of all, we note that the digit 2 is certainly visible whenever the hour includes a 2. This is the case for the hours 02, 12, 20, 21, 22 and 23, or $\frac{6}{24} = \frac{1}{4}$ of the hours. During the other $1 - \frac{1}{4} = \frac{3}{4}$ of the hours, it is visible whenever the tens-digit is a 2 (that is, from minutes 20 to 29, for a total of 10 minutes), or when the minutes digits read 02, 12, 32, 42 or 52. This gives us a total of 15 such minutes, which is $\frac{15}{60} = \frac{1}{4}$ of the 60 minutes in an hour. Altogether, this means that the digit 2 is visible somewhere on the display $\frac{1}{4} + \frac{1}{4} \times \frac{3}{4} = \frac{7}{16}$ of the time.

Solution for Problem 11.5:

If a division can be done without leaving a remainder, the divisor is certainly not greater than the dividend. This means that times with the property we are looking at must always have minutes less than (or equal to) the hours. From 8:30 to 8:59, this is certainly never the case. We can therefore focus our attention on the divisors of 9, 10 and 11. We first note that $9 = 3^2$ is a perfect square, with only the divisors 1, 3 and 9. All of these are between 0 and 59, and there are therefore three times from 9:00 to 9:59 when the minutes divide the hours without leaving a remainder, namely 9:01, 9:03 and 9:09. Next, $10 = 2 \times 5$ is the product of two different primes, and the divisors of 10 are therefore 1, 2, 5 and 10. Once again, all of these correspond to appropriate times, namely 10:01, 10:02, 10:05 and 10:10. Finally, 11 is a prime, with only the divisors 1 and 11. Since both 11:01 and 11:11 are between 11:00 and 11:30, this yields two more times with the required property, for a total of $3 + 4 + 2 = 9$ such times.

Solution for Problem 11.6:
Since the red watch is a minute fast each hour and the blue watch is a half a minute slow each hour, the times displayed on the two watches drift apart by $1\frac{1}{2} = \frac{3}{2}$ minutes each hour. Since the times they display in the morning are 30 minutes apart, the time that has elapsed from the time they were set until then is equal to $30 \div \frac{3}{2} = 30 \times \frac{2}{3} = 20$ hours. The red watch has gained 20 minutes in those 20 hours, and the actual time the watches are being checked in the morning is therefore 6:10, which means that the time they were set, which was 20 hours before that, was 10:10.

Solution for Problem 11.7:
If we take a look at the sum $1 + 2 + 3 + \cdots + 10 + 11 + 12$, we note that the sum $1 + 12$ of the first and last numbers in this expression is 13. This is also the sum $2 + 11$ of the second and next to last numbers, the sum $3 + 10$ of the third number from the left and the third number from the right, and so on. Since there are 12 numbers to be added altogether, there must be $12 \div 2 = 6$ such pairs that each add up to 13, and the sum of all the numbers is therefore equal to $6 \times 13 = 78$.

Solution for Problem 11.8:
Any problem dealing with days of the week can be expressed in terms of modular arithmetic. If you are familiar with this topic, what follows will likely seem quite obvious. However, even if you have never heard of the concept before, you will find that it is actually quite easy to apply; just a counting method that goes around in circles, really.

Let's start by replacing the names of the days of the week with roman numerals. We write I instead of Monday, II for Tuesday, III for Wednesday, IV for Thursday, V for Friday, VI for Saturday and VII for Sunday. In this notation, adding one day to Sunday does not give us a higher number, but since a Sunday is followed by the next Monday, the VII is followed by another I. Similarly, the day two days after VII is II and the day three days after VII is III. Actually, if we want to know the numeral of any day after a given one, we can simply add the number of days it should be later and then eliminate as many sevens as necessary (one seven for each complete week), until we obtain a numeral from I through VII. For instance, the day five days after VI (that is to say, the day five days after Saturday) is $VI + 5 = 11$, which yields $11 - 7 = 4$, or Thursday.

Now, if we want to know the day on which February 13th will fall in a year in which January the 13th is on a Friday, we simply note that January has 31 days, and since February 13th is therefore 31 days after Friday, January the 13th, we obtain $V + 31 = 36$, which yields $36 - 5 \times 7 = 1$, or Monday.

Knowing this, we can simply continue checking the days of the week for all the following 13th days of the month until we reach another Friday.

Since February has 28 days, March 13th is on $I + 28 = 29$, and $29 - 4 \times 7 = 1$ yields Monday.

Since March has 31 days, April 13th is on $I + 31 = 32$, and $32 - 4 \times 7 = 4$ yields Thursday.

Since April has 30 days, May 13th is on $IV + 30 = 34$, and $34 - 4 \times 7 = 6$ yields Saturday.

Since May has 31 days, June 13th is on $VI + 31 = 37$, and $37 - 5 \times 7 = 2$ yields Tuesday.

Since June has 30 days, July 13th is on $II + 30 = 32$, and $32 - 4 \times 7 = 4$ yields Thursday.

Since July has 31 days, August 13th is on $IV + 31 = 35$, and $35 - 4 \times 7 = 7$ yields Sunday.

Since August has 31 days, September 13th is on $VII + 31 = 38$, and $38 - 5 \times 7 = 3$ yields Wednesday.

Finally, since September has 30 days, October 13th is on $III + 30 = 33$, and $33 - 4 \times 7 = 5$ yields Friday, and we are done.

Solution for Problem 11.9:

May and July each have 31 days, while June has 30. This means that these three months have a total of 92 days. Since $92 = 13 \times 7 + 1$, the first day of May and the last day of July fall on the same day of the week. If this is a Friday, there are 14 Fridays in the three months, but in all other cases, there are certainly exactly 13. If July 31st is a Friday, August 13th must be a Thursday.

We see that August 13th can be on any day of the week, except Thursday, in a year with the given property.

Solution for Problem 11.10:

July and August each have 31 days, while September has 30. The three months therefore have a total of 92 days. If there are 14 Sundays in this period, the number of days from the first to the last of these is equal to $13 \times 7 + 1 = 92$. This implies that the first and last days of the period in question must certainly be Sundays.

We can now count backwards. Since July 1st is a Sunday, June 30th is a Saturday, and therefore, so is June 2nd (since $30 - 4 \times 7 = 2$). Continuing in this vein, June 1st is a Friday and May 31st is a Thursday, as is May 3rd (since $31 - 4 \times 7 = 3$). From this, we see that May 2nd is a Wednesday, May 1st is a Tuesday and April 30th is a Monday, as is April 2nd (since $30 - 4 \times 7 = 2$). We therefore see that April 1st is a Sunday, March 31st is a Saturday, and March 28th is therefore a Wednesday, and we are finished.

Solution for Problem 11.11:

The first two digits of 1991 are 19. The next two palindromic years must therefore start with the next two two-digit numbers, which are 20 and 21. This means that the next two palindromic years after 1991 are 2002 and 2112, and the number of years between these two is equal to $2112 - 2002 = 110$.

Solution for Problem 11.12:

The largest number smaller than 999 composed of all identical digits is 888, and twice this number is 1776. Going in the other direction, we note that there is no three-digit number larger than 999 composed of all identical digits, and we must therefore find the smallest four-digit number composed of all identical digits. Since this digit cannot be 0, the smallest digit in such a number must be 1, and we therefore see that 1111 is the required number, and twice this number is, of course, 2222.

Solution for Problem 11.13:

If Karl is now k years old and Anna is a years old, Karl was Anna's age $k - a$ years ago. Since his mother was 32 years old $k - a$ years ago, her current age is $32 + (k - a)$. We know that Karl, his mother and his sister Anna are now 56 years old in total, and this can be written as

$$k + 32 + (k - a) + a = 56,$$

which is equivalent to $2k = 24$ or $k = 12$. We see that Karl is 12 years old. Note that we do not have enough information to determine how old Anna or his mother are!

Solution for Problem 11.14:

Christian obviously wasn't born in 2000, since he would then be 1 and not $2 + 0 + 0 + 0 = 2$. If he was born in the 1990s, say in the year $199x$, his age would be $1 + 9 + 9 + x = 19 + x$. No matter what the value of x is, this number is certainly greater than 19, but since he was born after 1990, he can be at most 11, which yields a contradiction.

If Christian was born in the 1980s, say in the year $198x$, his age would be $1 + 9 + 8 + x = 18 + x$. The year $18 + x$ years after $198x$ is $1980 + x + 18 + x = 1998 + 2x$, and this cannot be 2001, as this would mean $1998 + 2x = 2001$, which is equivalent to $2x = 3$, which is impossible for integer values of x.

If Christian was born in the 1970s, say in the year $197x$, his age would be $1 + 9 + 7 + x = 17 + x$. The year $17 + x$ years after $197x$ is $1970 + x + 17 + x = 1987 + 2x$. If this is equal to 2001, we have $1987 + 2x = 2001$, which is equivalent to $2x = 14$ or $x = 7$. If Christian was born in 1977, he would have been $2001 - 1977 = 24$ in 2001, and $1 + 9 + 7 + 7 = 24$ does indeed hold.

Is this the only possible solution? What is Christian was born before 1970? In this case, he would be at least $2001 - 1970 = 31$ years old in 2001. The maximum sum of the digits in a year before 1970 is in the year 1899 with $1 + 8 + 9 + 9 = 27$, and therefore less than 31. We see that there can be no other solution to the problem, and Christian is therefore certainly 24 years old in 2001.

Solution for Problem 11.15:

If we add all three given sums, we obtain twice the sum of all three men's ages, as each will have been included twice. This means that

$109 + 56 + 85 = 250$ is twice the sum of all their ages and 125 is therefore the sum of all their ages. Since the sum of the ages of Joseph and René is 85, Gerhard's age must therefore be $125 - 85 = 40$.

If you find this explanation difficult to follow, try writing it out using variables. If we let j denote Joseph's age, g Gerhard's age and r René's age, we are given

$$g + j = 109,$$

$$g + r = 56,$$

$$r + j = 85.$$

Adding all three equations gives us

$$(g + j) + (g + r) + (r + j) = 109 + 56 + 85,$$

or

$$2(g + j + r) = 250.$$

This yields $g + j + r = 125$, and since we have $r + j = 85$, we obtain

$$g = (g + j + r) - (r + j) = 125 - 85 = 40.$$

Solution for Problem 11.16:

Let d denote Mr. Dreher's age in years and w his wife's age in years. When Mr. Dreher's wife is as old as he is now, he will be $d - w$ years older than he is now, and therefore he will be $d + (d - w) = 2d - w$ years old. Since his wife will then be d years old, the first sentence can be written as

$$(2d - w) + d = 108 \quad \text{or} \quad 3d - w = 108.$$

When Mr. Dreher was the age his wife is now, i.e. $d - w$ years ago, his wife was $w - (d - w) = 2w - d$ years old. Since he is now twice that age, we also have

$$d = 2(2w - d) \quad \text{or} \quad 3d = 4w.$$

Substituting in the first equation gives us

$$4w - w = 108 \quad \text{or} \quad 3w = 108 \quad \text{or} \quad w = 36,$$

and therefore

$$d = \frac{4}{3} \times w = \frac{4}{3} \times 36 = 48.$$

We can easily check that these numbers are correct. When Mr. Dreher's wife is as old as he is now, i.e. 48, Mr. Dreher will be 60, and the sum of their ages will indeed be $60 + 48 = 108$. When Mr. Dreher was his wife's current age, i.e. 36, his wife was 24, and he is now indeed twice that age, since $2 \times 24 = 48$.

Solution for Problem 11.17:

First of all, let us take a look at the prime decomposition of 660.

$$660 = 2 \times 2 \times 3 \times 5 \times 11.$$

Since none of Mr. Hofer's children is older than 20, one must be 11 years old, as any multiple of 11 would be too big. Since we already know that none of the children can be one year old (if this were the case, the product of the ages of the youngest and oldest child could not be the age of Mr. Hofer's mother, as it would be the age of one of his children), the youngest of the children must be at least 2.

Now, we know that one of the children is 4 years younger than another. If one of these two is the 11 year old, the other would be 7 or 15. Since 7 is not a factor of 660, this leaves only the option 15 for the age of the oldest child. For the youngest we now have two options, since we have two twos left. Either there is a third child, whose age is $2 \times 2 = 4$, or there are twins who are both 2 years old. In the first case, the product of the ages of the youngest and oldest child is $4 \times 15 = 60$, and this is certainly a possible solution, since Mr. Hofer's mother could certainly be 60 years old. In the second case, the product of the ages of the youngest and oldest child is $2 \times 15 = 30$, and this is not a possible solution, since Mr. Hofer's mother could not be 30 years old if she has a 15-year-old grandchild. The only possible solution resulting in this case therefore yields three children, whose ages are 4, 11 and 15, respectively.

Is there another possible solution? We have not yet considered the case in which the two children whose ages differ by four do not include the 11 year old. Due to the fact that we have two twos in the prime decomposition, this is only possible if both ages are even and one child is $2 \times 3 = 6$ years old and the other is $2 \times 5 = 10$ years old. These numbers, namely 6 and 10, also differ by 4. In this case, the product of the ages of the youngest and the oldest child is $6 \times 11 = 66$, and this is certainly a possible age for Mr. Hofer's mother. We see that this case yields a second solution, again with three children, but this time with the ages 6, 10 and 11.

Solution for Problem 11.18:
Since a minute has 60 seconds, an hour has 60 minutes, a day has 24 hours and a year has (about) 365 days, the number of seconds in a year is equal to

$$60 \times 60 \times 24 \times 365.$$

Of course, it is not difficult to do the multiplication, multiply the result by your age in years, and then round off. That does involve quite a bit of calculation though, or at the very least the use of a calculator. There is a much easier way to find the answer to this question, though.

Let's write the multiplication in a slightly more advantageous way. We can write

$$60 \times 60 \times 24 \times 365 = (6 \times 10) \times (6 \times 10) \times 24 \times 365$$
$$= 10 \times 10 \times (6 \times 365) \times (6 \times 24),$$

and this makes it much easier to figure the approximate value. The tens at the beginning simply deliver zeros at the end of the product, so we don't need to worry about those at all for now. The product 6×365 is approximately equal to 2000, and the product 6×24 is approximately equal to 150. Multiplying these two therefore yields 300,000, since $2 \times 15 = 30$, and we simply count the number of zeros in the other numbers and add them up. Recalling the two tens we had not included yet, we get an approximate value of 30,000,000 (or a 3 followed by seven zeros) for the number of seconds in a year. Assuming now that someone taking part in a math competition is likely to be somewhere between 10 and 18 years old, multiplying this by a number in this range will add a zero at the end (as the factor is greater than 10), and raise the value of the leading digit a bit from 3 (with 5 being a reasonable option). The number of seconds in a competition participant's life is therefore approximately 500,000,000 (or a 5 followed by eight zeros; i.e. 500 million).

Solution for Problem 11.19:
In the solution to the last problem, we showed that a year has approximately 30,000,000, or 30 million, seconds. If Tony Milliard's age in seconds is equal to 1 billion, or 1000 million, we can easily divide 1000 by 30, which gives a result somewhat larger than 30. We see that his age will be somewhere between 20 and 35, and the correct answer is therefore C.

Solution for Problem 11.20:
Let x denote the number of participants. The easiest way to approach this problem is to calculate the total times of all participants in two ways. On the one hand, we know that the average time of all the participants was 12.39 seconds, and since there were x participants, their total time was $12.39 \times x$ seconds. On the other hand, we similarly know that the total time of the three medallists was 10.5×3, and the total time of the $x - 3$ non-medallists was $13.2 \times (x - 3)$. Since the total time of the whole group is the sum of these two values, we have two different expressions, each of which gives us the total times of all participants, and this gives us the equation

$$12.39 \times x = 10.5 \times 3 + 13.2 \times (x - 3).$$

This is equivalent to

$$12.39 \times x = 31.5 + 13.2 \times x - 39.6 \quad \text{or} \quad 0.81 \times x = 8.1,$$

and dividing this equation by 0.81 gives us $x = 10$. We see that there were 10 participants in the race altogether.

Solution for Problem 11.21:
This problem is worth taking a really close look at. A lot of people get this one wrong, since it seems so obvious that the answer will simply be 20 minutes. Unfortunately however, it isn't. How can we argue this rigorously and properly understand what is going on?

Let us assume that the trams depart every x minutes and travel at a speed of t km per minute. Furthermore, let us assume that Christoph is riding his bike at a speed of b km per minute.

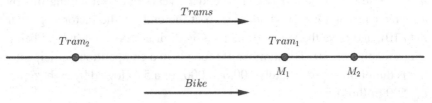

First of all, let us take a look at the situation in which the trams and the bike are traveling in the same direction. In the figure, both are going from left to right, as symbolized by the arrows. There are two trams on the track, $Tram_1$ and $Tram_2$, and the distance between them is equal to $t \times x$ km, as they departed x minutes apart and are both traveling at a speed of t km/min. M_1 denotes the spot in which Christoph is passed by $Tram_1$, and M_2 denotes the spot in which he is passed by $Tram_2$. Since 22 minutes pass from the time that Christoph is at M_1 and the time he is at M_2, the distance between M_1 and M_2 is equal to $22b$. On the other hand, in those same 22 minutes, $Tram_2$ travels from its spot through M_1 to M_2. This means that the distance $Tram_2$ travels in 22 minutes, i.e. $22t$, is equal to the distance between the two trams plus the distance Christoph travels in his bike in the same time, i.e. $tx + 22b$. This gives us the equation

$$22t = tx + 22b.$$

Now, let us consider the opposite situation.

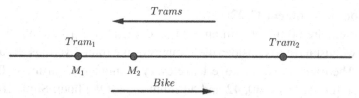

Now, the trams and the bike are traveling in opposite directions. In the figure, Christoph is going from left to right, while the trams are going from right to left, as symbolized once again by the arrows. Once again, there are two trams on the track, $Tram_1$ and $Tram_2$, and the distance between them is again equal to $t \times x$ km, as before. Also as before, M_1 denotes the spot in which Christoph meets $Tram_1$, and M_2 the spot in which he meets $Tram_2$. Since 18 minutes pass from the time that Christoph is at M_1 and the time he is at M_2, the distance between M_1 and M_2 is now equal to $18b$. On the other hand, in those same 18 minutes, $Tram_2$ travels from its spot to M_2, and this means that the distance $Tram_2$ travels in 18 minutes, i.e. $18t$, is now equal to the distance between the two trams minus the distance Christoph travels in his bike in the same time, i.e. $tx - 18b$. This therefore gives us the second equation

$$18t = tx - 18b.$$

Isolating tx in each of these equations, we therefore obtain

$$tx = 22t - 22b = 18t + 18b.$$

Simplifying the equation $22t - 22b = 18t + 18b$ yields $4t = 40b$ or $t = 10b$, and substituting for t in either the equation $22t = tx + 22b$ or $18t = tx - 18b$ yields

$$198b = 10bx \quad \text{or} \quad 19.8 = x.$$

We therefore see that the trams depart in intervals of 19.8 minutes, or 19 minutes and 48 seconds.

This is quite close to 20 minutes, of course, but not quite the same. This is a great example of an apparently easy mathematical problem turning out not to be at all as easy as it seems. Still, when you think about it, the actual solution isn't all that complicated either. It just turns out to be really important not to jump to any premature conclusions, if you want to find the correct answer. The key is to think it through rigorously from start to finish.

Solution for Problem 11.22:
Since there are 60 minutes in an hour, and 12, 15 and 10 are all divisors of 60, we just have to consider the minutes of the trains' various departure times. The trains to Krilatskoye leave every 12 minutes (starting at 0:02), and therefore at 6, 18, 30, 42 and 54 minutes past the hour. Similarly, the trains to Praschskaya leave every 15 minutes (starting at 0:03), and therefore at 3, 18, 33 and 48 minutes past the hour and the trains to Jugo-Sapadnaya leave every 10 minutes (starting at 0:02), and therefore at 2, 12, 22, 32, 42 and 52 minutes past the hour.

Listing all of these departure times in order, we see that a train leaves at

$$2, 3, 6, 12, 18, 22, 30, 32, 33, 42, 48, 52 \text{ and } 54$$

minutes past the hour. The longest interval between departure times (including the interval from 54 minutes past the hour to 2 minutes past the next hour) is from 33 to 42 minutes past the hour, i.e. 9 minutes long.

The longest possible interval that Joe may have to wait for the next train to leave is therefore 9 minutes.

Solution for Problem 11.23:
If x hours have passed since midnight, there are $24 - x$ hours left in the day. The traveler is told: "Until the end of the day, there still remain three

times two-ninths of the time that has already passed from the beginning of the day". This means that the $24 - x$ hours left in the day are equal to three times two ninths of x, and expressing this as an equation gives us

$$24 - x = 3 \times \frac{2}{9} \times x,$$

or $24 = \frac{5}{3} \times x$, which gives us $x = \frac{3}{5} \times 24 = \frac{72}{5} = 14\frac{2}{5}$. Since two-fifths of an hour is equal to 24 minutes, the time right now is 14:24. The 14:45 train therefore does not leave for another 21 minutes and there is plenty of time for the traveler to reach the train.

Solution for Problem 11.24:
Since Eugene has to help out 6 hours a week on average for 7 weeks, he must help out at least $6 \times 7 = 42$ hours altogether in that time. Since he helps out $8 + 3 + 6 + 7 + 7 = 31$ hours in the first five weeks, he must help out for at least another $42 - 31 = 11$ hours in the last 2 weeks. On average, that means another 5½ hours a week for the last two weeks.

Solution for Problem 11.25:
Since planes leave in the three directions every 40 minutes, every 60 minutes and every 72 minutes, respectively, there was one leaving at every multiple of these numbers before 12:00. The last time there was a plane leaving for all three destinations at once was therefore the least common multiple of minutes of these three numbers before 12:00. Since

$$\text{lcm}(40, 60, 72) = \text{lcm}(2^3 \times 5, 2^2 \times 3 \times 5, 2^3 \times 3^2) = 2^3 \times 3^2 \times 5 = 360,$$

this was 360 minutes, or 6 hours, earlier, and therefore at exactly 6:00.

Solution for Problem 11.26:
We know that it is 6 in the morning in Kumberg when it is midnight in Washington. This means that uncle Barack can be reached after he finishes work at 21:00 in Washington, or 3:00 in the morning in Kumberg, since this is 3 hours earlier. Similarly, he can be reached until he leaves for work at 8:30 in the morning in Washington, which is 14:30 in the afternoon in Kumberg (i.e. also 6 hours later). This means that Uncle Barack is reachable from 3:00 in the morning until 14:30, Kumberg time. Since Noah cannot call before 13:20, he can therefore only call Uncle Barack between 13:20 and 14:30, Kumberg time, which gives him a window of 70 minutes.

Solution for Problem 11.27:
While the first patient is being treated, the other four all have to wait. Similarly, three have to wait while the second is being treated, two have to wait while the third is being treated and one has to wait while the fourth is being treated. Since no one must wait while the last is being treated, it is best to treat the patient last whose treatment takes the longest. Working backward, it is then best to treat the patient fourth whose treatment takes the next longest, and so on, until finally the patient is treated first, whose treatment is the fastest. In this way, the total waiting time is equal to

$$4 \times 5 + 3 \times 7 + 2 \times 10 + 12 = 73 \text{ minutes.}$$

Solution for Problem 11.28:
Paula arrives at the hollow tree every 12 minutes. In other words, she arrives there at every multiple of 12 minutes, and since Niklas arrives there at every multiple of 10 minutes, they always arrive there simultaneously at common multiples of 12 and 10. The first time they meet there simultaneously is therefore after the least common multiple of 12 and 10, and since $12 = 2 \times 6$ and $10 = 2 \times 5$, we have lcm $(12, 10) = 2 \times 6 \times 5 = 60$. We see that they will run for exactly 60 minutes until they meet at the hollow tree and stop.

Solution for Problem 11.29:
A full red – red/yellow – green – green blink – yellow cycle lasts

$$40 + 3 + 45 + 4 + 3 = 95 \text{ seconds.}$$

The 30 minutes from 7 a.m. to 7:30 a.m. are equal to $30 \times 60 = 1800$ seconds, and ordinary division gives us

$$1800 = 1710 + 90 = 95 \times 18 + 90.$$

At exactly 7:30, we see that we are exactly 90 seconds after the light has just changed to red, and the light is therefore blinking green.

Solution for Problem 11.30:
A full green – yellow – red cycle lasts

$$50 + 5 + 30 = 85 \text{ seconds.}$$

Since the light changes to green at exactly 7 a.m., the number of times it changes to green between 7 a.m. and 7 p.m. is equal to the number of times the 85 second interval fits completely into that 12 hour interval. Since 12 hours are equal to

$$12 \times 60 \times 60 = 43200 \text{ seconds},$$

and

$$43200 \div 85 \approx 508.2.$$

We see that the light changes to green exactly 508 times between 7 a.m. and 7 p.m.

Chapter 12

Letter Puzzles and Digit Puzzles

All the problems in this book are mathematical puzzles, but there are some that look more like the type of puzzle you might see in a typical puzzle book or a newspaper. The problems presented in this chapter are of this type. You probably wouldn't be too surprised to see one of these alongside the crossword and the Sudoku in your daily paper.

PROBLEMS

Letter Puzzles

In each of the following puzzles, the same letters always represent equal digits and different letters always represent different digits. The first digit of a multi-digit number can never be 0.

Problem 12.1:
Determine the value of **E**.

$$
\begin{array}{r}
E\ R \\
+\ E\ R \\
+\ R\ E \\
\hline
S\ E\ E
\end{array}
$$

(O-95-5)

Problem 12.2:
Determine the value of **C**.

$$
\begin{array}{ccc}
  \text{A} & \text{B} & \text{C} \\
+ & \text{A} & \text{B} \\
\hline
3 & 2 & 5
\end{array}
$$

(S5K-07-3)

Problem 12.3:
Determine the smallest possible value of **X**.

$$\text{X} = \text{F} \times \text{E} \times \text{R} \times \text{I} \times \text{E} \times \text{N}$$

(S5K-08-2)

Problem 12.4:
Determine the value of the three-digit number **ABC**.

$$
\begin{array}{ccc}
  \text{A} & \text{B} & \text{C} \\
+ & \text{A} & \text{B} \\
+ & & \text{A} \\
\hline
4 & 7 & 4
\end{array}
$$

(S5K-12-10)

Problem 12.5:
Determine the smallest possible value of the four-digit number **ZWEI**.

$$
\begin{array}{cccc}
   & \text{E} & \text{I} & \text{N} \\
+ & \text{E} & \text{I} & \text{N} \\
\hline
\text{Z} & \text{W} & \text{E} & \text{I}
\end{array}
$$

(S5K-14-9)

Problem 12.6:
In this sum, the letters **X**, **Y** and **Z** represent digits, while the other digits are given. Determine the value of $\text{X} \times \text{Y} \times \text{Z}$.

$$
\begin{array}{cccc}
   & \text{X} & \text{Y} & \text{Z} \\
+ & \text{Y} & \text{Z} & 0 \\
+ & \text{Z} & 0 & 0 \\
\hline
1 & 7 & 1 & 8
\end{array}
$$

(S5K-15-10)

Problem 12.7:
Determine the value of **X**.

$$X952 \div 1X = X04$$

(StU-09-A3)

Problem 12.8:
A, B, C and **D** are four different single digit numbers. The product of **A, B** and **C** is 20. **D** is a multiple both of **A** and **B**, but not of **C**. the sum of **A, C** and **D** is 17. Determine the four-digit number

ABCD.

(S5K-16-6)

Number Placement Puzzles and Magic Squares

In each of the following puzzles, the object is to place numbers in some type of two-dimensional grid according to a given set of rules.

Problem 12.9:
Each of the squares in the figure is to be filled with one of the digits from 1 to 9. Every digit is to be used somewhere and the sum of the digits in each row and each column is to be 13. Determine a distribution of the digits fulfilling these conditions.

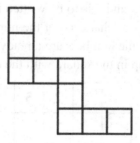

(O-96-5)

Problem 12.10:
Write the numbers 1, 2, 3, 4 and 5 in the squares in such a way that both the sum of the numbers in the horizontal row and the sum of the numbers in

the vertical column equal 8. Which number must be in the center square?

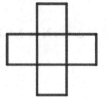

(S5K-05-3)

Problem 12.11:

A table is divided into six cells as shown. Is it possible to place the six numbers 1, 2, 3, 4, 5 and 6 into the cells, such that the sum of the numbers in each row is the same? Is it possible to place them such that the sums in each column are the same?

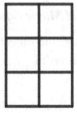

(StU-96-B3)

Problem 12.12:

One of the numbers 2, 3, 5 and 7 is to be written in each of the squares of the given array in such a way that each of the numbers is written in every column and every row. Some numbers are already written in their squares. Which number will end up in the square with the question mark?

		2	5
		7	
			2
?	3		

(S5K-09-1)

Problem 12.13:

In the given magic square, the sums of the numbers in each row, in each column, and in each of the two diagonals are equal. This sum is called the "magic constant" of the magic square. Determine this magic constant.

21	8	
		20

(S5K-98-6)

Problem 12.14:

In the magic square shown in the figure, the numbers from 1 to 16 are arranged in such a way that the sum of the four numbers in each row, each column and each diagonal is equal to 34. Which number must be written in the square marked X?

	X	3	
		15	12
16	9	4	
	11		

(S5K-17-3)

Problem 12.15:

The numbers 1, 2, 3, . . . , 9 are to be placed in the circles of the diagram in such a way that the sum of any three numbers connected by a straight line is always equal. Which number must be written in the circle labeled A?

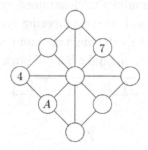

(StU-16-B2)

Problem 12.16:
The numbers 1, 2, 3, 4, 5 and 6 are to be placed in the circles of the diagram in such a way that the sum of the numbers in the three circles along each side of the triangle is the same. Which number cannot be placed in the lower left corner, if the number 2 is placed at the top of the triangle?

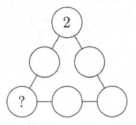

(StU-06-A3)

Problem 12.17:
Numbers are to be written in the cells of the pyramid in such a way that each number is the sum of the two numbers immediately below it. Which number must be written at the top of the pyramid?

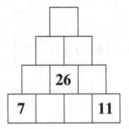

(StU-96-A7)

Problem 12.18:
A number is to be written on each brick of the wall shown on the right. The numbers on the lower bricks are determined by the numbers on the two bricks directly above them. If two neighboring bricks have the numbers a and b, the brick under them will obtain the number $a + b + ab$, as shown. Which number will be written on the bottom brick?

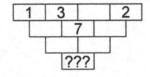

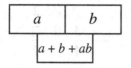

(StU-13-A9)

Missing Digits

Problem 12.19:
In the multiplication

$$3\,\square\,\square \times 1\,\square = \square\,3\,2\,\square$$

the same digit goes in each rectangle to yield a correct result. Which digit is it?

(S5K-03-5)

Problem 12.20:
In the expression

$$98\,x \,\underline{}\,\underline{} + \boxed{} = 2011$$

digits are to be placed on each line to create a correct equation. Which two-digit number must be placed in the gray box?

(S5K-11-2)

Problem 12.21:
This year's Schul5Kampf takes place on the first of July, i.e. 1.7.2015. I would therefore like to write the year 2015 in the following form:

$$5 \times (17 \times [\;][\;] + [\;][\;]) = 2015.$$

Each of the brackets stands for a digit, so that we have two two-digit numbers in the expression. The second (right-hand) two-digit number is smaller than the first (left-hand) one. Determine the missing right-hand two-digit number.

(S5K-15-2)

Problem 12.22:
Determine the sum of the three missing digits.

$$\begin{array}{r} \square\ 8\ 2 \\ -\ 3\ 8\ \square \\ \hline 4\ \square\ 7 \end{array}$$

(StU-98-A3)

Problem 12.23:
Determine the sum of the three missing digits.

$$\begin{array}{r} 9\ 3\ \square \\ +\ \square\ 6\ \square \\ \hline 1\ 1\ 0\ 1 \end{array}$$

(StU-05-A3)

A Bonus 3D Problem

Problem 12.24:
A positive integer is written in each vertex of a quadratic pyramid in such a way that each of the five sides of the pyramid (4 triangles and 1 square) yield the same result if the numbers in all their corners are multiplied. Two of the numbers are given in the figure. (D ... 2; S ... 10). Determine all five numbers.

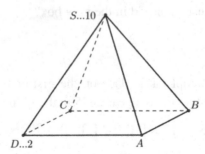

(StU-15-B2)

TIPS

Tip for Problem 12.1:
The values of **R** and **S** should be obvious right away. Once you know them, the value of **E** should follow pretty quickly.

Tip for Problem 12.2:
There are only two possible values for **A** from the start. Think about the options for the other two digits in each case.

Tip for Problem 12.3:
Can 0 be one of the digits? Note also that one of the digits turns up twice in the product.

Tip for Problem 12.4:
Take another look at the tip from Problem 12.2.

Tip for Problem 12.5:
The smallest four-digit number is 1000. You can get pretty close to that!

Tip for Problem 12.6:
Try working backwards. Once you get the value of **Z**, you will find **Y** and then **X**.

Tip for Problem 12.7:
Check out the possible values of the ones-digits first.

Tip for Problem 12.8:
There is only one way to express 20 as the product of three different single-digit numbers.

Tip for Problem 12.9:
If four sums are all to be 13, the total of all these sums is 4×13. This is almost the sum of all the numbers, but not quite. What else do we need to consider?

Tip for Problem 12.10:
This is very similar to the last one. The same idea should get you through quite quickly.

Tip for Problem 12.11:
Add up all of the numbers as a first step. How many groups of numbers do you need to fill each of the rows? How many for each of the columns?

Tip for Problem 12.12:
Don't go straight for the solution, but just play around with this first. Which of the empty squares can you fill with the appropriate number right away? Taking it step by step will get you there!

Tip for Problem 12.13:
Try introducing a variable for some value you don't know, but would like to know in order to answer the question.

Tip for Problem 12.14:
The same idea that worked in Problem 12.12 will work again here. If you look at the grid, you will see that you know a lot more already than might be immediately obvious.

Tip for Problem 12.15:
It may not seem like it at first, but if you look at this carefully, you will see that it is really another magic square. Also, if you read carefully, you will see that you already know the magic constant.

Tip for Problem 12.16:
You can solve this by trying out all five of the other numbers and checking which ones you will be able to complete the puzzle with. There is a faster way to figure this out, though, if you notice what happens when you add all three sums together.

Tip for Problem 12.17:
Try introducing a variable for the number in one of the bottom cells.

Tip for Problem 12.18:
Just take this one step by step. Nothing to it.

Tip for Problem 12.19:
You could just try out all the possibilities, but that would take a while. Take a look at the ones-digits. What do you notice there? Which digits does that immediately eliminate from consideration?

Tip for Problem 12.20:
This is just a division problem with remainder.

Tip for Problem 12.21:
Take another look at Problem 12.20. This one is almost the same.

Tip for Problem 12.22:
Since we are subtracting, don't forget to work from right to left!

Tip for Problem 12.23:
You won't be able to find the missing ones-digits, but their sum should be pretty easy to figure out.

Tip for Problem 12.24:
Once again, it will be useful to introduce a variable for one of the unknown numbers. One is enough, though!

SOLUTIONS

Solution for Problem 12.1:
Since **R** is the first digit of **RE**, its value is certainly not 0. Looking at the ones-digits in the addition, we see that **R** must therefore be equal to 5, since $2 \times \mathbf{R} + \mathbf{E}$ could not otherwise end in **E**. Furthermore, the value of **S** must be either 1 or 2, since it is the result of a carryover from the tens-column to the hundreds-column, and $2 \times \mathbf{E} + \mathbf{R}$ cannot be larger than $2 \times 9 + 5 = 23$.

$$
\begin{array}{r}
\mathbf{E}\ \mathbf{R} \\
+\ \mathbf{E}\ \mathbf{R} \\
+\ \mathbf{R}\ \mathbf{E} \\
\hline
\mathbf{S}\ \mathbf{E}\ \mathbf{E}
\end{array}
$$

The addition must therefore either be

$$
\begin{array}{r}
\mathbf{E}\ 5 \\
+\ \mathbf{E}\ 5 \\
+\ 5\ \mathbf{E} \\
\hline
1\ \mathbf{E}\ \mathbf{E}
\end{array}
\quad \text{or} \quad
\begin{array}{r}
\mathbf{E}\ 5 \\
+\ \mathbf{E}\ 5 \\
+\ 5\ \mathbf{E} \\
\hline
2\ \mathbf{E}\ \mathbf{E}
\end{array}
$$

Since there is certainly a carryover of 1 from the ones-column to the tens-column, the addition in the tens-column must therefore be either

$$\mathbf{E} + \mathbf{E} + 5 + 1 = 10 + \mathbf{E} \quad \text{or} \quad \mathbf{E} + \mathbf{E} + 5 + 1 = 20 + \mathbf{E}.$$

The first of these options simplifies to $\mathbf{E} = 4$, while the second simplifies to $\mathbf{E} = 14$. Since **E** must be a digit, the value of **E** is therefore 4, with the addition being

$$
\begin{array}{r}
4\ 5 \\
+\ 4\ 5 \\
+\ 5\ 4 \\
\hline
1\ 4\ 4
\end{array}
$$

Solution for Problem 12.2:

Since the first digit of the result here is 3, the value of **A** must be either 2 or 3, depending on whether there is a carryover or not.

$$
\begin{array}{ccc}
 & \text{A} & \text{B} & \text{C} \\
+ & & \text{A} & \text{B} \\
\hline
 & 3 & 2 & 5
\end{array}
$$

We note, however, that there must also be a carryover from the tens-column if **A** is equal to 3, since the result in the tens-column is 2, and therefore less than 3. This means that the value of **A** is certainly 2, and the addition can be written as

$$
\begin{array}{ccc}
 & 2 & \text{B} & \text{C} \\
+ & & 2 & \text{B} \\
\hline
 & 3 & 2 & 5
\end{array}
$$

Next, we note that the value of **B** must therefore be 9, since the result in the tens-column could otherwise not equal 2. This means that the value of C must be 6, and we obtain

$$
\begin{array}{ccc}
 & 2 & 9 & 6 \\
+ & & 2 & 9 \\
\hline
 & 3 & 2 & 5
\end{array}
$$

with the value of C being 6.

Solution for Problem 12.3:

We first note that none of the digits can be 0, as the value of **X** would then also be 0, and not all different letters would then represent different digits. Taking a closer look at the factors in the product

$$\mathbf{X} = \mathbf{F} \times \mathbf{E} \times \mathbf{R} \times \mathbf{I} \times \mathbf{E} \times \mathbf{N},$$

we note that the factor **E** turns up twice, while all other factors are only included once. This means that **X** will be minimal if **E** = 1, and the letters **F, R, I** and **N** represent the next larger digits, namely 2, 3, 4 and 5, in some order. The value of X in this case is then

$$2 \times 1 \times 3 \times 4 \times 1 \times 5 = 120.$$

Solution for Problem 12.4:

We see from the hundreds-column that the value of **A** must be either 3 or 4, and from the tens-column that the value of **B** + **A** must be either 6 or 7.

$$
\begin{array}{c c c}
\text{A} & \text{B} & \text{C} \\
+\ \text{A} & \text{B} \\
+ & & \text{A} \\
\hline
4 & 7 & 4
\end{array}
$$

(Since the value of **A** is either 3 or 4, the value of **B** + **A** cannot be greater than 10.) From the ones-column, we see that **C** + **B** + **A** is therefore equal to 14, and since this produces a carryover into the tens-column, we now know that **B** + **A** must be equal to 6. This means that the value of **C** is 8 and the value of **A** is 4, since there is no carryover from the tens-column to the hundreds-column. Since the value of **B** + **A** is 6, we see that the value of **B** is therefore 2, and the addition is

$$
\begin{array}{c c c}
4 & 2 & 8 \\
+\ 4 & 2 \\
+ & & 4 \\
\hline
4 & 7 & 4
\end{array}
$$

with the value of **ABC** being 428.

Solution for Problem 12.5:

The number **ZWEI** is certainly a four-digit number, and the value of **Z** is certainly 1, as it only results as a carryover from the hundreds-digit.

$$
\begin{array}{c c c c}
 & \text{E} & \text{I} & \text{N} \\
+ & \text{E} & \text{I} & \text{N} \\
\hline
\text{Z} & \text{W} & \text{E} & \text{I}
\end{array}
$$

We would like the value of **ZWEI** to be as small as possible, and so we can try to find values of the digits, such that the value of **W** is 0. This is, indeed, possible.

If **E** + **E** = 10 in the hundreds-column (with no carry-over), we have **E** = 5. This means that the sum in the tens-column must be **I** + **I** = 4, with a carryover of 1 from the ones-digit, and thus **I** = 2. This is only possible for $N = 6$ (and therefore $N + N = 12$), and the smallest possible value of **ZWEI** is therefore 1052 with the addition being

$$
\begin{array}{r}
5 \; 2 \; 6 \\
+ \; 5 \; 2 \; 6 \\
\hline
1 \; 0 \; 5 \; 2
\end{array}
$$

Solution for Problem 12.6:
Taking a look at the ones-column, we get $Z + 0 + 0 = 8$, and thus $Z = 8$.

$$
\begin{array}{r}
X \; Y \; Z \\
+ \; Y \; Z \; 0 \\
+ \; Z \; 0 \; 0 \\
\hline
1 \; 7 \; 1 \; 8
\end{array}
$$

Moving on to the tens-column, we then get $Y + Z + 0 = 11$ (the value of this sum is obviously not 1, since $Z = 8$), and therefore $Y = 3$. As a final step, the hundreds-column therefore yields $X + Y + Z + 1 = 17$, or $X + 3 + 8 + 1 = 17$, which gives us $X = 5$. The complete addition is therefore

$$
\begin{array}{r}
5 \; 3 \; 8 \\
+ \; 3 \; 8 \; 0 \\
+ \; 8 \; 0 \; 0 \\
\hline
1 \; 7 \; 1 \; 8
\end{array}
$$

and the required value of $X \times Y \times Z$ is therefore $5 \times 3 \times 8 = 120$.

Solution for Problem 12.7:
Taking a look at the ones-digits in this division, we see that the number $4 \times X$ must end in the digit 2.

$$X952 \div 1X = X04$$

This means that the value of X must be either 3 or 8. It is now easy to check the two options. Since $X = 8$ would give us $804 \times 18 = 14{,}432$ (and not 8952), the only remaining option is $X = 3$, and

$$3952 \div 13 = 304$$

is indeed a correct equation.

Solution for Problem 12.8:
The only way to express 20 as the product of three different single-digit numbers is as

$$20 = 1 \times 4 \times 5.$$

The digits **A**, **B** and **C** are therefore 1, 4 and 5 in some order. Since **D** is a multiple of **A** and **B**, but not of **C**, and 5 has no single-digit multiples, we definitely have **C** = 5, and **D** is a multiple of 1 and 4. The value of **D** must therefore be 8. Finally, since **A** + **C** + **D** = 17, we have **A** + 5 + 8 = 17, or **A** = 4. Since this implies **B** = 1, the number **ABCD** is therefore 4158.

Solution for Problem 12.9:
The key squares in solving this problem are the ones marked with *a* and *b* in the following figure:

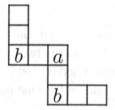

Since the sum of the numbers in all four rows and columns is to be 13, the total of all these sums is $4 \times 13 = 52$. This total sum includes all numbers from 1 to 9 at least once, but the numbers in the *a* and *b* squares twice. Noting that the sum of all numbers from 1 to 9 is equal to

$$1 + 2 + 3 + \cdots + 9 = 45.$$

We see that the sum of the three numbers in these squares must be equal to $52 - 45 = 7$. The only way 7 can be the sum of three different numbers of these is as follows $1 + 2 + 4$. These are therefore the numbers in the squares marked *a* and *b*.

   If *a* is either 1 or 2, the sum $a + b + ?$ will be $1 + 2 + ?$ in either the row or column ending in *a*, but this is not possible, since the number in the third square would then be 10. It therefore follows that the number in square a must be 4, and the numbers in the two squares marked b must be 1 and 2 in either order, as shown.

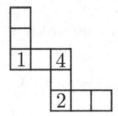

Since the sum in each row and column is 13, this immediately gives us the numbers between these squares:

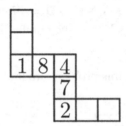

This leaves the digits 3, 5, 6 and 9 unused, and since the two numbers above the 1 must sum to $13 - 1 = 12$, and the two to the right of the 2 must sum to $13 - 2 = 11$, this means that 3 and 9 must lie above 1 in some order and 5 and 6 to the right of 2 in some order. One possible solution is therefore the following:

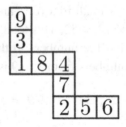

Solution for Problem 12.10:

This is almost the same as the last problem, really. The sum in both the row and the column is 8, and the sum of both is therefore $8 + 8 = 16$. This is also the sum of all the numbers from 1 through 5, with the number in the middle counted twice. Since

$$1 + 2 + 3 + 4 + 5 = 15,$$

the number in the center square must be $16 - 15 = 1$.

Solution for Problem 12.11:

The sum of all the numbers from 1 to 6 is

$$1 + 2 + 3 + 4 + 5 + 6 = 21.$$

In order for the sums in each of the three rows to be the same, these numbers have to be divided into three groups of two numbers each, with the sum in each pair being equal to $21 \div 3 = 7$. This is possible, as we can use the three pairs. (1, 6), (2, 5) and (3, 4). One possible placement in the table would therefore be

1	6
2	5
3	4

On the other hand, for the sums in the two columns to be the same, the numbers would have to be divided into two groups of three numbers each, with the sum in each group of three being equal to $21 \div 2$, which is not an integer. Such a placement is therefore not possible.

Solution for Problem 12.12:

It is best to go at this step by step. In each of the steps illustrated, the bold numbers are forced because of the other numbers already present in the square's row and column.

		2	5
		7	
			2
?	**3**	**5**	

		2	5
		7	
		3	**2**
?	**3**	**5**	

Now, let us look at the right-hand column. The digit 7 must be placed somewhere in this column, but it cannot be placed in the square marked **X**, as there is already a 7 in this square's row.

		2	5
		7	X
		3	2
?	3	5	

This means that the 7 must be placed in the bottom right

		2	5
		7	
		3	2
?	3	5	**7**

and the only remaining option for the square with the question mark is therefore 2, as all other digits are already present in the bottom row.

		2	5
		7	
		3	2
2	3	5	7

Note that we can fill up the square completely in a unique way as

3	7	2	5
5	2	7	3
7	5	3	2
2	3	5	7

but this is not required to answer the question as posed.

Solution for Problem 12.13:

If we knew all the numbers in the top row, we could simply add them up to obtain the magic constant. Since we don't know this number, let us call it x.

21	8	X
		20

The value of the magic constant is then $21 + 8 + x = 29 + x$. If we can find the value of x, we are finished. In order to do this, we can fill in some if the values in the other empty squares, dependent on the unknown x. Since the magic constant is $29 + x$, this must be the sum of the three numbers in the right-hand column, and the bottom square must therefore contain the number $(29 + x) - 20 - x = 9$.

21	8	x
		20
		9

Similarly, looking at the main diagonal (from top left to bottom right), the center square must contain $(29 + x) - 21 - 9 = x - 1$.

21	8	x
	x–1	20
		9

Also, considering the other diagonal, the bottom left square must contain $(29 + x) - x - (x - 1) = 30 - x$.

21	8	x
	x–1	20
30–x		9

We now have two different ways to calculate the missing values in each of the two remaining empty squares. Let us take a look at the middle square of the bottom row. Since it is the third number in the middle column. Along with 8 and $x - 1$, its value must be $(29 + x) - 8 - (x - 1) = 22$. On the other hand, since it is also the third number in the bottom row along with $30 - x$ and 9, its value must be $(29 + x) - (30 - x) - 9 = -10 + 2x$. This gives us the equation

$$22 = -10 + 2x,$$

which is equivalent to $x = 16$. The magic constant is therefore equal to

$$29 + 16 = 45.$$

Note that we can now fill in the entire magic square to obtain

21	8	16
10	15	20
14	22	9

Solution for Problem 12.14:
We are given that the sums of the numbers on all rows, columns and diagonals are 34. In a first step, we can immediately add the numbers $34 - 16 - 9 - 4 = 5$ and $34 - 3 - 15 - 9 = 7$

	X		3
		15	12
16	9	4	5
7	11		

and then $34-3-12-5 = 14, 34-7-11-14 = 2$ and $34-15-4-2 = 13$.

	X	13	3
		15	12
16	9	4	5
7	11	2	14

The upper left square must therefore contain the number $34 - x - 13 - 3 = 18 - x$, and the square immediately below the x must contain the number $34 - x - 9 - 11 = 14 - x$.

18-x	**X**	**13**	**3**
	14-x	**15**	**12**
16	**9**	**4**	**5**
7	**11**	**2**	**14**

Since the sum in the diagonal from top-left to bottom-right must also be 34, we obtain the equation

$$(18 - x) + (14 - x) + 4 + 14 = 34,$$

which reduces to $x = 8$. In fact, we can now complete the whole magic square:

10	**8**	**13**	**3**
1	**6**	**15**	**12**
16	**9**	**4**	**5**
7	**11**	**2**	**14**

Solution for Problem 12.15:
Since the numbers from 1 to 9 are to be placed in the 9 circles, and the three sums going downwards from left to right are all equal, these must be equal to

$$(1 + 2 + 3 + \cdots + 9) \div 3 = 45 \div 3 = 15.$$

This means that 15 is the sum of any three numbers in a straight line. We can therefore attack this problem in the same way as we did the last few, by simply filling in what we know, step by step, until we get some information about the value of **A**.

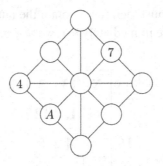

First of all, since $15 - 7 - A = 8 - A$, the middle circle must contain the number $8 - A$.

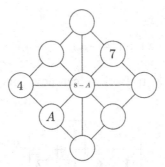

Furthermore, since $15 - 4 - (8 - A) = A + 3$, the circle on the right must contain the number $A + 3$.

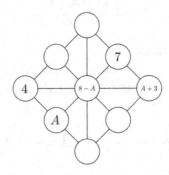

Since $15 - 7 - (A + 3) = 5 - A$ and $15 - 4 - A = 11 - A$, the number in the top circle must be $5 - A$ and the number in the bottom circle must be $11 - A$.

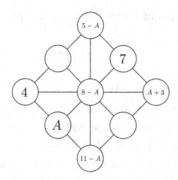

Since the three numbers in the vertical diagonal of the figure must also add up to 15, this gives us the equation

$$(5 - \mathbf{A}) + (8 - \mathbf{A}) + (11 - \mathbf{A}) = 15,$$

and this is equivalent to $\mathbf{A} = 3$.

As was the case in the previous problems, we can fill in the complete puzzle to obtain

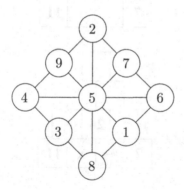

Solution for Problem 12.16:

If we put the numbers from 1 to 6 into the circles, the sum of all the numbers in the circles is $1 + 2 + 3 + 4 + 5 + 6 = 21$. Furthermore, if we let s stand for the sum of the three numbers in three circles along any of the three sides of the triangle, we see that $s + s + s$ must be equal to $21 + 2 + x + y$, where 2, x and y are the three numbers in the corners, as these are each used in two of the sums. This means that we obtain the equation

$$3s = 23 + x + y.$$

Since we can also write this as $3s = 3 \times 7 + (2 + x + y)$, we see that x and y must be chosen in such a way that $2 + x + y$ is divisible by 3. This is not possible if we choose 5 as the value of x, as none of the numbers $2 + 5 + 1 = 8, 2 + 5 + 3 = 10, 2 + 5 + 4 = 11$ and $2 + 5 + 6 = 13$ is divisible by 3. We therefore see that 5 cannot be placed in the lower left corner if all of the conditions are to be met.

Solution for Problem 12.17:

If we place a variable x in one of the two bottom cells, the pyramid can be completed step by step in the following way:

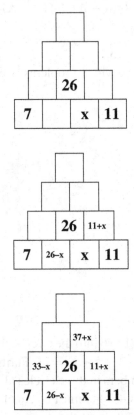

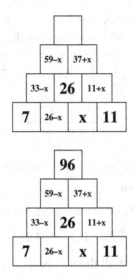

We see that 96 must be the number in the uppermost cell, no matter which value we choose for x.

Solution for Problem 12.18:

We are given the fact that the number in each of the bricks is equal to the sum of the two numbers in the bricks above it together with their product. Since $1 + 3 + 1 \times 3 = 7$, we see that the number under the 1 and the 3 must be 7, as shown.

1	3	x	2
	7	7	

Introducing the variable x for the unknown number between the 3 and the 2 in the upper row gives us $3 + x + 3x = 7$ for the center brick, and this is equivalent to $x = 1$.

1	3	1	2
	7	7	

Because of $1 + 2 + 1 \times 2 = 5$, $7 + 7 + 7 \times 7 = 63$, $7 + 5 + 7 \times 5 = 47$ and $63 + 47 + 63 \times 47 = 3071$, we can complete the wall as

1	3	1	2
	7	7	5
		63	47
			3071

and note that the number written on the bottom brick will be 3071.

Solution for Problem 12.19:

If we look at the ones-digits of the numbers involved in the equation

$$3\square\square \times 1\square = \square 3\ 2\square$$

we see that the ones-digits of all the numbers are the same. For most of the digits, this will not be the case. For instance, $2 \times 2 = 4$ does not end in 2 and $8 \times 8 = 64$ does not end in 8. The only digits we need to consider are therefore 0, 1, 5 and 6. We can immediately forget about 0, as the first digit of the result of the multiplication cannot be 0, but it can also not be 1, as the multiplication of a number larger than 300 with a number larger than 10 certainly yields a number larger than 3000, whose first digit cannot be 1. The only digits we need to check are therefore 5 and 6. Since $366 \times 16 = 5856$ (and not 6326), we can also eliminate the digit 6 from contention, leaving only 5, and since

$$355 \times 15 = 5325$$

does indeed hold, we see that 5 is the digit we are looking for.

Solution for Problem 12.20:

This is really just an abstract way to write a division problem. Looking at the numbers

$$98 \times \underline{\quad\quad} + \underline{\quad\quad} = 2011$$

we can simply divide 2011 by 98, and obtain 20 with a remainder of 51. In other words, we can write

$$98 \times 20 + 51 = 2011.$$

The number we are looking for is therefore 51.

Solution for Problem 12.21:

This is really another division problem. There is a preliminary step, however, since the given equation

$$5 \times (17 \times [\ ][\ ] + [\ ][\ ]) = 2015$$

simplifies to

$$17 \times [\ ][\ ] + [\ ][\ ] = 403.$$

Dividing 403 by 17, we obtain 23 and a remainder of 12. In other words, we can write

$$17 \times 23 + 12 = 403$$

or

$$5 \times (17 \times 23 + 12) = 2015,$$

and the number we are looking for is therefore 12.

Solution for Problem 12.22:

Working from right to left, we first notice that $12 - 5 = 7$, and the missing ones-digit must therefore be 5, yielding

$$
\begin{array}{r}
\square\ 8\ 2 \\
-\ 3\ 8\ 5 \\
\hline
4\ \square\ 7
\end{array}
$$

We have used a carry-over from the tens-digit. And the missing tens-digit therefore result from $17 - 8 = 9$, yielding

$$
\begin{array}{r}
\square\ 8\ 2 \\
-\ 3\ 8\ 5 \\
\hline
4\ 9\ 7
\end{array}
$$

and since there was also a carryover from the hundreds-digit, $7 - 3 = 4$ means that the missing hundreds-digit must be 8, yielding

$$
\begin{array}{r}
8\ 8\ 2 \\
-\ 3\ 8\ 5 \\
\hline
4\ 9\ 7
\end{array}
$$

The required sum of the missing digits is therefore $5 + 9 + 8 = 22$.

Solution for Problem 12.23:
Since there is a one in the ones-digit of the result of the addition, the sum of
the two missing ones-digits must be either 1 or 11. Noting that $3 + 6 = 9$,
and not 10, we see that there is a carryover from the ones- to the tens-column,
and the sum of the two missing ones-digits is therefore certainly 11.

$$
\begin{array}{r}
9 \ 3 \ \square \\
+ \ \square \ 6 \ \square \\
\hline
1 \ 1 \ 0 \ 1
\end{array}
$$

Since there is also a carry-over from the tens-column, we have
$9 + 1 + 1 = 11$ in the hundreds-column, and the third missing digit
is therefore 1. The sum of the three missing digits is therefore $11 + 1 = 12$.

Solution for Problem 12.24:

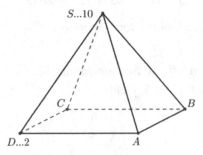

If we let a represent the number to be written in **A**, we see that the product
of the numbers written in the corners of **SDA** is $10 \times 2 \times a = 20a$. Since
the product of the numbers in the corners of **SAB** is also $20a$, the number
written in **B** must be $20a \div (10a) = 2$, and similarly for **SCD**, the number in
C must be $20a \div 20 = a$. From this we see that the product of the numbers
written in the corners of the base **ABCD** must also equal $20a$, and this yields
the equation

$$2 \times a \times 2 \times a = 20a.$$

Since the value of a is given as positive, it is certainly not 0, and the equation
therefore simplifies to $a = 5$, and we see that the numbers written in **A** and
C are both 5.

Chapter 13

How Far, How Fast, How Long

A lot of people find puzzles dealing with speeds and rates of change to be much harder than other types of problems. One reason for this may be the fact that such problems inevitably involve multiple intertwined units of measurement, like miles per hour combined with miles and hours. This certainly forces a certain level of concentration in dealing with such questions, but that doesn't mean that they can't be a lot of fun to take on! Try your hand at these.

PROBLEMS

Problem 13.1:
Tobias can eat 2 chocolate bars in 10 minutes and Lena can eat one in 20 minutes. How long will it take them both to eat 3 bars together, if they both eat at their usual speeds?

(StU-02-A7)

Problem 13.2:
Farmer Hiasl can mow the whole of his meadow by himself in 5 hours. If Miatzl helps him, they can do the job together in 3 hours. How long would it take Miatzl to do the job by herself?

(O-97-3)

Problem 13.3:
Four caterpillars eat four leaves in 4 minutes. How long does it take 20 caterpillars to eat 100 leaves?

(S5K-01-7)

Problem 13.4:
We know that 5 cats can catch 5 mice in 5 minutes. How many cats can catch 100 mice in 100 minutes?

(StU-93-11)

Problem 13.5:
18 bag gluers glue 3240 bags in 1½ hours. How many bag gluers glue 2160 bags in 2 hours?

(StU-99-A8)

Problem 13.6:
A passenger train takes 8 minutes and 30 seconds to pass through a tunnel. A freight train takes 11 minutes to pass through the same tunnel. The passenger train is going at a speed 5 m/s faster than the freight train. How long is the tunnel?

(S5K-01-5)

Problem 13.7:
A 120 m long train is crossing a 300 m long bridge at a speed of 72 km/h. How many seconds pass from the moment the front of the train drives onto the bridge until the end of the train leaves the bridge?

(StU-04-B1)

Problem 13.8:
Stefan and Christoph are running in the woods with their dog Rex. Stefan runs the 3 km course in 15 minutes. It takes Christoph 20 minutes to do the same. The whole time they are running, Rex runs back and forth between them at a pace that is twice as fast as Stefan's. How far does Rex run altogether?

(S5K-00-4)

Problem 13.9:
Klaus runs twice as fast as he walks. On his way to school, he runs two-thirds of the time and walks one-third of the time, and this takes him a total of 16 minutes. How long will it take him to get home if he walks two-thirds of the time and runs one-third of the time?

(StU-95-12)

Problem 13.10:
Two brothers, Markus and Thomas, both walk to school along the same path. Markus runs half the *way* at a speed of 12 km/h and walks half the *way* at a speed of 4 km/h. Thomas walks half the *time* and runs half the *time*, and he also runs three times as fast as he walks. Both take the same time to get to school. How fast is Thomas?

(StU-98-B6)

Problem 13.11:
It normally takes Ms. Habergeiss 25 minutes to drive the 30 km from her home to her work. What is her average speed during the drive? How many minutes will she be late if she drives the first half of the way with 60 km/h instead of her usual speed? What should then be her average speed for the second half of the way if she still wants to arrive at work on time?

(StU-00-B2)

Problem 13.12:
A pool has two input valves and one drain. The whole pool can be filled through the left valve in 20 minutes. Using only the right valve, it takes 30 minutes to fill the pool. The whole pool can be emptied through the drain in 15 Minutes. By accident, both valves and the drain are opened simultaneously. How long will it take until the pool is filled? Or can it not be filled at all under these circumstances?

(S5K-02-9)

Problem 13.13:
On the evening of March 14th, farmer Huber determines that the hay in his loft would last him 30 days if he only feeds his cows. If he only feeds his sheep, the hay will last him for 70 days. Of course, he has to feed both his cows and his sheep. On the evening of which day will he run out of hay if he does not put any of his animals on a diet?

(StU-02-B6)

TIPS

Tip for Problem 13.1:
There are several ways you could think about this. Maybe the simplest is to consider how many chocolate bars (or, in other words, which fraction of a chocolate bar) each of them eats in a minute.

Tip for Problem 13.2:
Note that you already know what fraction of the meadow Farmer Hiasl can mow in an hour and you would like to know which fraction of the meadow Miatzl could mow by herself in an hour.

Tip for Problem 13.3:
How much will 20 caterpillars eat in 4 minutes?

Tip for Problem 13.4:
If you don't read this carefully, you might think that this is the same question as the last one. But take a closer look. This one is actually even easier.

Tip for Problem 13.5:
If you have been doing these problems in order, you will probably be detecting a pattern by now. It's always a good idea to reduce something to a unit with problems like these. How many bags does this group glue per hour?

Tip for Problem 13.6:
There are several things we don't know here. We don't know how long the tunnel is, nor how fast either of the trains are going. We can assume that the trains are of the same length, as there is nothing said about that, but we will have to introduce at least one variable for one of the things we don't know; maybe even more than one.

Tip for Problem 13.7:
Be careful! How far does the train have to travel altogether?

Tip for Problem 13.8:
How long is Rex running? How fast is he running?

Tip for Problem 13.9:
Once again, it will help to introduce variables here. We don't know how far the school is from his home, and we don't know his walking or running speeds. Be careful that the units you introduce match up!

Tip for Problem 13.10:
Take another look at the tip for the last problem. This will help you here, as well.

Tip for Problem 13.11:
This is a pretty straightforward calculation as long as you watch out for your units.

Tip for Problem 13.12:
What fraction of the pool is being filled by each of the input valves per minute? What fraction is being emptied by the drain per minute?

Tip for Problem 13.13:
This is a lot like the last one when you think about it a bit. What fraction of the hay do the cows require each day? What fraction of the hay do the sheep require?

SOLUTIONS

Solution for Problem 13.1:
If Tobias eats 2 chocolate bars in minutes, he eats $\frac{2}{10} = \frac{1}{5}$ of a chocolate bar in one minute. Similarly, if Lena eats a chocolate bar in 20 minutes, she eats $\frac{1}{20}$ of a chocolate bar in one minute. Adding these two numbers, we see that they eat

$$\frac{1}{5} + \frac{1}{20} = \frac{4}{20} + \frac{1}{20} = \frac{5}{20} = \frac{1}{4}$$

of a chocolate bar in a minute together. This means that it takes them 4 minutes to eat one chocolate bar together, or $3 \times 4 = 12$ minutes to eat three chocolate bars together.

Solution for Problem 13.2:
Farmer Hiasl can mow $\frac{1}{5}$ of the meadow by himself in an hour. If it takes Miatzl x hours to do this by herself, this means that she can mow $\frac{1}{x}$ of the meadow by herself in an hour. We are given that they can mow the meadow together in 3 hours, and this means that they can mow $\frac{1}{3}$ of the meadow in

a hour if they work together. This gives us the equation

$$\frac{1}{5} + \frac{1}{x} = \frac{1}{3},$$

which is equivalent to $\frac{1}{x} = \frac{1}{3} - \frac{1}{5} = \frac{5}{15} - \frac{3}{15} = \frac{2}{15}$. In other words, we have $x = \frac{15}{2} = 7\frac{1}{2}$, and we see that it would take Miatzl 7½ hours to mow the whole meadow by herself.

Solution for Problem 13.3:
Since $20 = 4 \times 5$, we know that 20 caterpillars can eat $4 \times 5 = 20$ leaves in 4 minutes. Since $100 \div 20 = 5$, we see that 20 caterpillars will eat 100 leaves in $4 \times 5 = 20$ minutes.

Solution for Problem 13.4:
It doesn't really matter how many cats there are. If we know that some number of cats catch 5 mice in 5 minutes, those same cats catch a mouse a minute on average. This means that those same 5 cats can catch 100 mice in 100 minutes.

Solution for Problem 13.5:
We can do this in two steps. If 18 bag gluers glue 3240 bags in 1½ $= \frac{3}{2}$ hours, they glue $3240 \div \frac{3}{2} = 3240 \times \frac{2}{3} = 2160$ bags per hour. This means that they glue twice as many in 2 hours, but it also means that half that number can glue 2160 bags in 2 hours, and we see that this requires $18 \div 2 = 9$ bag gluers.

Solution for Problem 13.6:
If we introduce a few strategic variables, this one won't be difficult at all. Let t denote the length of the tunnel in meters and f the speed of the freight train in meters per second. This means that the speed of the passenger train is $f + 5$ meters per second. Since the freight train passes through the tunnel in 11 minutes, or $60 \times 11 = 660$ seconds, we obtain

$$660 \times f = t.$$

Also, since the passenger train passes through the tunnel in 8 minutes and 30 seconds, or $60 \times 8 + 30 = 510$ seconds, we also obtain

$$510 \times (f + 5) = t.$$

Since both expressions are equal, we have

$$660f = 510(f + 5),$$

which simplifies to $150f = 2550$ or $f = 17$. The length of the tunnel is therefore

$$660 \times 17 = 11220\,\text{m} = 11.22\,\text{km}.$$

By the way, if you don't think a train tunnel can be that long, you obviously don't live in the Alps!

Solution for Problem 13.7:
This one isn't hard at all, but a bit tricky. You can still get it wrong if you don't read the question carefully. The total distance the train has to travel is equal to the length of the bridge (the distance the front of the train travels from the point in time when the train drives onto the bridge until the front of the train leaves the end of the bridge) plus the length of the train. This means that the train has to travel a distance of $300 + 120 = 420$ meters. At a speed of 72 km/h (or 72,000 meters per hour), that takes $\frac{420}{72000} = \frac{7}{1200}$ hours, or $\frac{7}{1200} \times 60 = \frac{7}{20}$ minutes or $\frac{7}{20} \times 60 = 21$ seconds.

Solution for Problem 13.8:
Since Rex is running for the whole time, he is running until both have finished, and this means that he is also running for 20 minutes, or one-third of an hour. We know that Stefan runs 3 km in 15 minutes, or $3 \times 4 = 12$ km per hour. Since Rex runs twice that fast, he runs at a speed of $12 \times 2 = 24$ km per hour. This means that he runs one-third that distance in 20 minutes, or $24 \div 3 = 8$ km altogether.

Solution for Problem 13.9:
Let d denote the distance from his home to his school in meters. Furthermore, let w denote his walking speed in meters per minute. His running speed is then equal to $2w$ meters per minute. Finally, let t denote the time it takes him to get home in minutes under the given assumptions.

If we first consider his trip to school, we know that he ran two-thirds of the 16 minutes of his total trip and walked the other third of the 16 minutes.

We therefore see that the distance d to school can be written in the form

$$d = \frac{2}{3} \times 16 \times 2w + \frac{1}{3} \times 16 \times w = \frac{5}{3} \times 16 \times w.$$

For his trip home, we know that he ran one-thirds of the t minutes of his total trip and walked the other two-thirds of the t minutes. We therefore see that the distance d to school can also be written in the form

$$d = \frac{1}{3} \times t \times 2w + \frac{2}{3} \times t \times w = \frac{4}{3} \times t \times w.$$

Since these two distances are equal, we obtain the equation

$$\frac{5}{3} \times 16 \times w = \frac{4}{3} \times t \times w,$$

which simplifies to

$$t = \frac{3}{4} \times \frac{5}{3} \times 16 = 20.$$

It therefore takes Klaus 20 minutes to get home under the given conditions.

Solution for Problem 13.10:
Let d denote the distance between the brothers' home and their school in km. Furthermore, let t denote the total time it takes the two to get to school. Also, let m_r denote the time that Markus spends running and m_w the time he spends walking. Finally, let t_w denote Thomas' walking speed, which implies that his running speed is $3 \times t_w$.

Since Markus walks half the way and runs half the way, we have

$$12 \times m_r = \frac{d}{2} = 4 \times m_w,$$

which implies $m_r = \frac{d}{24}$ hours and $m_w = \frac{d}{8}$ hours. The total time it takes Markus to get to school is therefore equal to

$$t = m_r + m_w = \frac{d}{24} + \frac{d}{8} = \frac{4d}{24} = \frac{d}{8}$$

hours. Since Thomas runs half the time and walks half the time, he spends $\frac{d}{8} \div 2 = \frac{d}{16}$ hours walking and the same amount of time running. Since he walks at a speed of t_w and runs at a speed of $3t_w$, the total distance he travels to school totals

$$t_w \times \frac{d}{12} + 3t_w \times \frac{d}{12} = d.$$

This is equivalent to $\frac{4}{12} \times t_w \times d = d$, which is equivalent to $t_w = 3$. We see that Thomas walks at a speed of $3\,\mathrm{km/h}$, while he runs at three times this speed, or $9\,\mathrm{km/h}$.

Solution for Problem 13.11:
If Ms. Habergeiss drives 30 km in 25 minutes, she drives

$$30 \times \frac{60}{25} = 72$$

km in an hour, and her average speed is therefore $72\,\mathrm{km/h}$.

If she drives the first half of the distance, i.e. the first $30 \div 2 = 15\,\mathrm{km}$, at a speed of $60\,\mathrm{km/h}$ it will take her $\frac{15}{60} = \frac{1}{4}$ of an hour $= 15$ minutes to travel this first distance, instead of $25/2 = 12\frac{1}{2}$ minutes, causing a delay of $2\frac{1}{2}$ minutes. If she wants to arrive at work on time, she has $25 - 15 = 10$ minutes, or $\frac{10}{60} = \frac{1}{6}$ of an hour left to travel the remaining $15\,\mathrm{km}$. If she wants to do this, she must travel this part at an average speed of $15 \div \frac{1}{6} = 15 \times 6 = 90\,\mathrm{km/h}$.

Solution for Problem 13.12:
The left valve fills $\frac{1}{20}$ of the pool every minute and the right valve fills $\frac{1}{30}$ of the pool every minute, while the drain empties $\frac{1}{15}$ of the pool every minute. If they are all opened simultaneously, the total gain per minute is equal to

$$\frac{1}{20} + \frac{1}{30} - \frac{1}{15} = \frac{3}{60} + \frac{2}{60} - \frac{4}{60} = \frac{1}{60}$$

of the pool every minute. This means that the pool will be completely full in exactly 60 minutes.

Solution for Problem 13.13:
Each day, farmer Huber's cows eat up $\frac{1}{30}$ of the hay. Similarly, each day, his sheep eat up $\frac{1}{70}$ of the hay. Together, their daily consumption uses up

$$\frac{1}{30} + \frac{1}{70} = \frac{7}{210} + \frac{3}{210} = \frac{10}{210} = \frac{1}{21}$$

of the hay. It follows that the hay will last exactly 21 days if all the animals eat it together, and it will therefore run out 21 days after March 14th, or on April 4th.

Chapter 14

Triangles

Most geometric questions can be reduced somehow to considerations of triangles. We might be asking about angles or lengths of line segments or areas of certain shapes; most anything can be chopped up into triangles in some way. Considering this, it isn't surprising to note that there are a lot of different kinds of fairly elementary questions we can ask about just triangles themselves. Here are some interesting ones. We have chosen these because they each have some special characteristic to separate them from the pack. You might find these a bit more challenging than the problems in other chapters, but that should make it all the more satisfying, when you get the solutions!

(Note that the symbol [XYZ] is used throughout to denote the area of a triangle XYZ.)

PROBLEMS

Problem 14.1:
Let ABC be a triangle with obtuse angle in A. Extend the side AB beyond A by the length of side $b = AC$ to a point D and beyond B by the length of side $a = BC$ to a point E. Give reasons why the area of triangle BCD must exceed the area of triangle BEC.

(StU-09-B6a)

Problem 14.2:
Points P, Q and R lie on the sides AB, BC and AC of a triangle ABC, respectively, such that PQ is parallel to AC, and PR is parallel to BC. The

triangle ABC has an area of $24\,\text{cm}^2$ and the triangle PBQ has an area of $6\,\text{cm}^2$. Determine the area of triangle PQR.

(StU-03-B5)

Problem 14.3:
In a rectangle $ABCD$ with area $3120\,\text{cm}^2$, M is the common point of the diagonals, E is the midpoint of BC, and F is the midpoint of MC. Determine the area of triangle AEF or give reasons, why the area cannot be uniquely determined from the information given.

(StU-99-A10)

Problem 14.4:
The lengths of the sides of a triangle are three consecutive prime numbers. The perimeter of the triangle is equal to 97. How long are the sides of the triangle?

(S5K-17-7)

Problem 14.5:
Circles k_A, k_B and k_C are drawn with midpoints in the vertices of a triangle ABC with perimeter $p = 32$ in such a way that each of the circles is externally tangent to the other two circles. The radius r_A of the circle k_A with midpoint A is equal to 5. How long is the side $a = BC$?

(S5K-08-3)

Problem 14.6:
We are given an acute triangle ABC. A circle k is drawn with diameter AB. Let k intersect BC in D and AC in E. Now, draw the diagonals of the quadrilateral $ABDE$. Their point of intersection is a special point with respect to triangle ABC. Which one? Justify your claim!

(StU-07-B7b)

Problem 14.7:
In an acute triangle ABC, the altitudes through A and B meet the opposite sides BC and AC in X and Y, respectively. The perpendicular bisector s of XY intersects AB in P. What special property does P have? Give reasons for your answer!

(StU-95-6)

Problem 14.8:
Let ABC be a right triangle with right angle in A and $AC = 12$. Points D and E lie on AB and BC, respectively, with DE perpendicular to BC. We know that $BD = 17$ and $BE = 15$. How long is AD?

(StU-02-B7)

Problem 14.9:
Let ABC be a right triangle with right angle in A. The lengths of its sides are $AB = 12$, and $AC = 5$. A circle k is tangent to AC in A and also tangent to BC. Calculate the radius of k.

(StU-07-B7A)

Problem 14.10:
Let ABC be a triangle with sides of lengths 6, 5 and 5. Calculate the radius of the circumcircle of ABC.

(StU-05-B7A)

Problem 14.11:
In a triangle, β is five times as large as α, and γ is six times as large as α. Determine β.

(S5K-03-4)

Problem 14.12:
We are given an isosceles triangle ABC with base AB and $\alpha = 20°$. A point D lies on the line segment AB such that $AD = AC$. Determine the angle contained by the line segments CD and CB.

(StU-95-3)

Problem 14.13:
The incircle of a triangle ABC with $\alpha = \angle BAC = 62°$, $\beta = \angle CBA = 46°$ is tangent to BC in D, to AC in E, and to AB in F. Determine the angle between the line segments DE and DF.

(StU-09-B3)

Problem 14.14:
In triangle ABC, we have $AB = AC$. The point D lies on the side AC, and the point E lies on the side AB in such a way, that CDE is equilateral,

and $\angle ECB = \angle BAC = \alpha$. Calculate the measure of α from the given information.

(S5K-06-9)

Problem 14.15:

Let ABC be a triangle with $\alpha = \angle BAC = 45°$ and $\beta = \angle CBA = 30°$. The point D lies on AB such that $AD = AC$, and E lies on BC such that $BE = BD$. Prove that the triangle DEC is isosceles.

(StU-02-B4)

Problem 14.16:

In a triangle ABC, CD is the internal bisector of $\angle ACB$, and BE is the internal bisector of $\angle ABC$. Let S be the point of intersection of CD and BE and let $\angle BAC = \alpha$ and $\angle BSD = 4\alpha$. Prove that the measure of α is uniquely determined by the given information and calculate the measure of α.

(StU-18-B2)

Problem 14.17:

In an isosceles triangle ABC with base AB, we are given $\gamma = \angle ACB = 98°$. The circle k with diameter BC intersects the internal bisector of $\beta = \angle CBA$ in B and a second time in P. Determine the angle contained by the line segments AC and PC.

(StU-08-A4)

Problem 14.18:

Let ABC be a triangle with obtuse angle in A. Extend the side AB beyond A by the length of side $b = AC$ to a point D and beyond B by the length of side $a = BC$ to a point E. It turns out that the triangle DAC is equilateral, and that BC bisects $\angle ACE$. Determine the angles in triangle ABC.

(StU-09-B6b)

Problem 14.19:

In a triangle ABC, we are given $\alpha = \angle BAC = 37°$ and $\beta = \angle CBA = 78°$. Let D denote the midpoint of AC, and let S denote the common point of the perpendicular bisector of AC and the internal bisector of β. Determine the angle subtended by the line segments SB and SD.

(S5K-04-8)

Problem 14.20:
Kevin claims that he has drawn a right-angled triangle ABC with hypotenuse AB, in which the internal bisector of the angle in A passes through the triangle's centroid S. Prove that Kevin must be wrong.

(StU-12-A7)

TIPS

Tip for Problem 14.1:
You can use the triangle inequality; in every triangle, each side is shorter than the sum of the other two sides.

Tip for Problem 14.2:
Since PQ is parallel to AC, triangles ABC and PBQ are homothetic with B as center of homothety.

Tip for Problem 14.3:
If two triangles have a common side, the ratio of their areas is equal to the ratio of their altitudes on the common side.

Tip for Problem 14.4:
The length of the shortest side of the triangle must be less than one-third of the triangle's perimeter; the length of the longest side must be more than one-third of the perimeter. This greatly reduces the number of possibilities that have to be checked.

Tip for Problem 14.5:
Express both the perimeter of ABC and the length of $a = BC$ in terms of the radii r_A, r_B and r_C.

Tip for Problem 14.6:
Note the right angles in the semi-circle!

Tip for Problem 14.7:
Just like you did in the last problem, note the right angles. There is a circle in play once again; you just need to find it!

Tip for Problem 14.8:
If corresponding angles are equal in two triangles, their third angles are also equal, and the triangles are similar. Together with the Pythagorean theorem, this should be enough to solve this problem.

Tip for Problem 14.9:
This is closely related to the previous problem. Let r denote the radius of k and try to set up an equation.

Tip for Problem 14.10:
ABC is an isosceles triangle and its circumcenter is therefore on its axis of symmetry.

Tip for Problem 14.11:
Establish an equation using the fact that the angles of any triangle sum to $180°$.

Tip for Problem 14.12:
In every triangle, the angles sum to $180°$. In addition, the angles at the base of an isosceles triangle are equal. Note that drawing line segments of equal length from a common point yields an isosceles triangle with vertices in the common starting point and the two endpoints of the segments.

Tip for Problem 14.13:
Note that the tangent distance from a point to a circle is uniquely determined. Triangles AEF, BDF and CDE are therefore isosceles.

Tip for Problem 14.14:
Focus on the angle in C.

Tip for Problem 14.15:
In order to prove that a triangle is isosceles, it is sufficient to show that two of its angles are equal. The proof becomes even easier if you pay attention to special characteristics of the quadrilateral $ADEC$.

Tip for Problem 14.16:
Concentrate on the sums of the angles in the triangles ABC and BCS and on $\angle BSD$.

Tip for Problem 14.17:
Note that BC is a diameter of the circle k. What do we know about angles in a semi-circle?

Tip for Problem 14.18:
The triangle BEC is isosceles with $\angle CBA$ as its external angle in B.

Tip for Problem 14.19:

Convince yourself that S lies outside ABC. Mark the common point P of the internal bisector of β with AC and determine the angles in triangle SDP.

Tip for Problem 14.20:

Note that, according to the information given, the incenter of triangle ABC lies on the median through vertex A. Show that this leads to a contradiction.

SOLUTIONS

Solution for Problem 14.1:

Note that the triangles BCD and BEC have a common altitude h_c in C. By the triangle inequality, we have

$$BD = AB + AD = AB + AC > BC = BE.$$

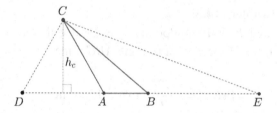

We therefore have

$$[BCD] = \frac{1}{2} \times BD \times h_c = \frac{1}{2} \times (AB + AC) \times h_c > \frac{1}{2} \times BE \times h_c = [BEC].$$

Solution for Problem 14.2:

ABC and PBQ are homothetic with respect to B, since PQ and AC are parallel. Note that, the ratio of corresponding areas is $a^2 : b^2$ if and only if the ratio of corresponding lengths is $a : b$. From

$$[ABC] : [PQR] = 24 : 6 = 4 : 1$$

it therefore follows that

$$AB : PB = BC : QB = AC : PQ = 2 : 1$$

holds. We therefore see that P, Q and R are the midpoints of the sides AB, BC and AC, respectively. From this, we see that the respective sides of all four triangles PQR, PBQ, PRA and RQC are half as long as the parallel sides of *ABC*, and all four triangles are therefore congruent. The area of *PQR* is therefore the same as the area of *PBQ*, i.e. $6\,\text{cm}^2$.

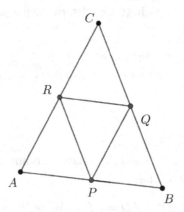

Solution for Problem 14.3:

We first note that *E* is the midpoint of *BC*, and it therefore follows that triangle *AEC* has half the area of triangle *ABC*. Since

$$[ABC] = \frac{3120}{2}\,\text{cm}^2 = 1560\,\text{cm}^2,$$

we therefore have $[AEC] = 780\,\text{cm}^2$. Since *F* is the midpoint of *MC* and the lengths of *AM* and *MC* are half the length of *AD*, we also have

$$AF = \frac{3}{4} \times AC,$$

and the area of *AEF* is three-quarters of the area of *AEC*, or $\frac{3}{4} \times 780 = 585\,\text{cm}^2$.

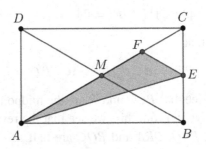

Solution for Problem 14.4:

The length of the shortest side of the triangle must be less than one-third of the triangle's perimeter, and the length of the longest side must be more than one third of the perimeter. The length of the shortest side must therefore be less than $\frac{97}{3} = 32\frac{1}{3}$, and the length of the longest side must exceed $32\frac{1}{3}$. The length of the second longest side of the triangle must therefore be one of the two prime numbers neighboring 32, that is, either 37 or 31.

In the first case, the side lengths would be 31, 37 and 41, giving a triangle with a perimeter greater than 97.

In the second case, we get side lengths 29, 31 and 37, yielding the required perimeter of 97.

Solution for Problem 14.5:

First, we note that

$$a = r_B + r_C, \ b = r_A + r_C, \ c = r_A + r_B,$$

and $p = 2(r_A + r_B + r_C)$ hold. We therefore obtain

$$a = r_B + r_C = \frac{p}{2} - r_A = 16 - 5 = 11.$$

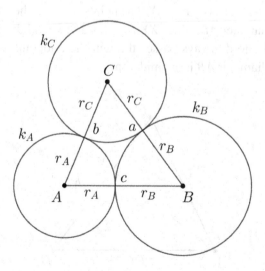

Solution for Problem 14.6:

Since the angle subtended by the diameter of a semi-circle on any point of the semi-circle is always 90° , and D and E both lie on k, we know that AD is perpendicular to BC, as is BE to CA. Points D and E are therefore the feet of the altitudes of ABC in A and B, respectively. The point of intersection of the diagonals AD and BE in $ABDE$ is therefore the orthocenter H of ABC.

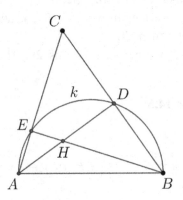

Solution for Problem 14.7:

Since X and Y are the feet of the altitudes from A and B, respectively, AX is perpendicular to BX, as is BY to AY. The points X and Y therefore lie on the circle k with diameter AB. Since XY is a chord of k, and the perpendicular bisector s of a chord always passes through the midpoint of the circle, s intersects the diameter AB in its midpoint.

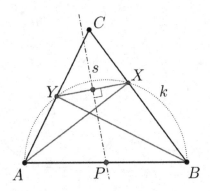

Solution for Problem 14.8:

Since *DE* is perpendicular to *BC*, *BED* is a right triangle with right angle in *E*. This triangle shares its angle in *B* with the right triangle *ABC*. Consequently, *ABC* and *EBD* are similar.

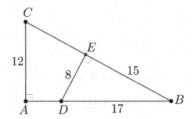

From the Pythagorean theorem, it follows that

$$DE = \sqrt{BD^2 - BE^2} = \sqrt{17^2 - 15^2} = 8.$$

Since *ABC* and *EBD* are similar, we have $AB : AC = BE : DE$, and we therefore obtain

$$AB = \frac{BE \times AC}{DE} = \frac{15 \times 12}{8} = 22.5,$$

and finally

$$AD = AB - BD = 22.5 - 17 = 5.5.$$

Solution for Problem 14.9:

According to the Pythagorean theorem, we have

$$BC = \sqrt{AB^2 + AC^2} = \sqrt{12^2 + 5^2} = 13.$$

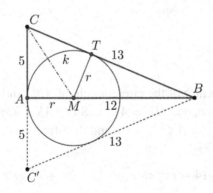

Let M denote the center of the circle k and let T denote the point of tangency of k and BC. Furthermore, let r denote the radius of k. The right triangles ABC and TBM have the angle in B in common, and are therefore similar with $MT:MB = AC:BC$. From $MB = AB - AM = 12 - r$, we then obtain

$$r:(12-r) = 5:13$$

$$\Leftrightarrow 13r = 5 \times (12-r)$$

$$\Leftrightarrow 18r = 60$$

$$\Leftrightarrow r = \frac{10}{3}.$$

Solution for Problem 14.10:

Let AB be the base of ABC. Then $AB = 6$ and the axis of symmetry of ABC intersects AB in its midpoint X. Note that AXC is a right triangle. Therefore, $XC = \sqrt{AC^2 - AX^2} = \sqrt{5^2 - 3^2} = 4$.

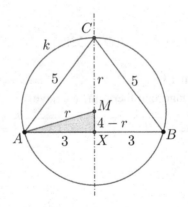

Let M denote the center of the circumcircle of ABC and r its radius. Then $UA = UB = UC = r$, and $UX = 4 - r$. Since $AX \perp XM$, the Pythagorean theorem yields $UA^2 = AX^2 + XU^2$ and therefore

$$r^2 = 3^2 + (4-r)^2 \Leftrightarrow r^2 = 25 - 8r + r^2 \Leftrightarrow r = \frac{25}{8}.$$

Solution for Problem 14.11:

We know that $\beta = 5\alpha$, $\gamma = 6\alpha$, and since the sum of the angles in a triangle is always equal to $180°$, we therefore have

$$\alpha + \beta + \gamma = \alpha + 5\alpha + 6\alpha = 12\alpha = 180°.$$

This is equivalent to

$$\alpha = \frac{180°}{12} = 15°.$$

It then follows that $\beta = 5\alpha = 75°$.

Solution for Problem 14.12:

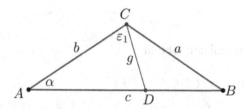

Since ABC is an isosceles triangle with base AB, we have

$$\angle CBA = \angle BAC = 20°$$

and

$$\angle ACB = 180° - 2 \times 20° = 140°.$$

Furthermore, we have $AD = AC$, and triangle ADC is therefore isosceles with base CD. From this, we obtain

$$\angle CDA = \angle ACD = \frac{180° - 20°}{2} = 80°.$$

Finally, we get

$$\angle DCB = \angle ACB - \angle ACD = 140° - 80° = 60°.$$

Solution for Problem 14.13:

Since *BDF* and *CDE* are isosceles triangles, we have

$$\angle CDE = \frac{180° - \gamma}{2} = 90° - \frac{\gamma}{2},$$

$$\angle FDB = \frac{180° - \beta}{2} = 90° - \frac{\beta}{2}.$$

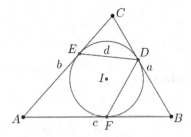

From this, we immediately obtain

$$\angle EDF = 180° - (\angle CDE + \angle FDB)$$

$$= 180° - \left(180° - \frac{\beta + \gamma}{2}\right) = \frac{\beta + \gamma}{2} = \frac{180° - \alpha}{2} = 59°.$$

Solution for Problem 14.14:

Since we are given $AB = AC$, we know that the triangle ABC is isosceles. We therefore have $\angle ACB = \angle CBA$. From $\angle DCE = \angle ACE = 60°$ and $\angle ECB = \angle BAC = \alpha$, we obtain

$$\angle CBA = \angle ACB = \angle ACE + \angle ECB = 60° + \alpha.$$

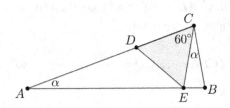

Considering the angles in *ABC*, we therefore obtain

$$\alpha + 2 \times (60° + \alpha) = 180° \Leftrightarrow 3\alpha + 120° = 180° \Leftrightarrow 3\alpha = 60°,$$

and finally, $\alpha = 20°$.

Solution for Problem 14.15:
From $\alpha = \angle BAC = 45°$ and $\beta = \angle CBA = 30°$, it follows that

$$\gamma = \angle ACB = \angle ACE$$
$$= 180° - (45° + 30°)$$
$$= 105°$$

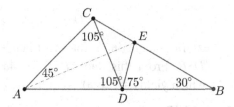

holds. From $BD = BE$ and $\beta = \angle CBA = 30°$, we obtain

$$\angle BDE = \angle DEB = \frac{180° - 30°}{2} = 75°,$$

and therefore

$$\angle EDA = 180° - \angle BDE = 180° - 75° = 105° = \angle ACE.$$

Furthermore, we also have $AD = AC$. This implies that the obtuse triangles *ADE* and *ACE* correspond in two side lengths and the obtuse angle opposite the common side *AE*. Since *AE* lies opposite the largest angle, it is the longest side in both triangles. Therefore, *ADE* and *ACE* are congruent, which yields $ED = EC$, and *DEC* is therefore isosceles.

Solution for Problem 14.16:

Let $\angle CBA = \beta$ and $\angle ACB = \gamma$. We know that

$$\alpha + \beta + \gamma = 180° \Leftrightarrow \beta + \gamma = 180° - \alpha$$

holds in every triangle, and we therefore have

$$\angle CBS + \angle SCB = \frac{\beta + \gamma}{2} = \frac{180° - \alpha}{2}$$

$$= 90° - \frac{\alpha}{2}.$$

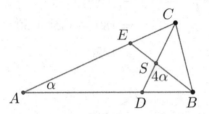

Since $\angle DSB$ is the exterior angle in S for the triangle BCS, we have $\angle CBS + \angle SCB = \angle DSB$. From this and $\angle BSD = 4\alpha$, it follows that $90° - \frac{\alpha}{2} = 4\alpha$ holds, which yields $\frac{9\alpha}{2} = 90°$, and finally $\alpha = 20°$.

Solution for Problem 14.17:

Since the triangle ABC is isosceles, we have

$$\alpha = \beta = \frac{180° - \gamma}{2} = \frac{180° - 98°}{2} = 41°.$$

Since BC is a diameter of k, we also have $\angle BPC = 90°$. This implies

$$\angle PCB = 90° - \frac{\beta}{2} = 69.5°,$$

and finally, we obtain

$$\angle ACP = \angle ACB - \angle PCB = 98° - 69.5° = 28.5°.$$

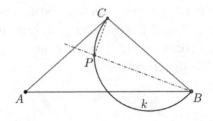

Solution for Problem 14.18:

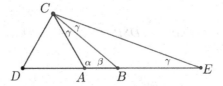

We are given that the triangle *DAC* is equilateral, and we therefore have $\angle CAD = 60°$ and $\alpha = \angle BAC = 120°$. We are also given that *BC* bisects $\angle ACE$, and that *BEC* is isosceles due to $BE = BC$. We therefore also have $\angle CEB = \angle BCE = \angle ACB = \gamma$. Furthermore, $\angle CBA$ is an external angle in *B* in the isosceles triangle *BEC*, and we therefore have $\beta = \angle CBA = \angle CEB + \angle BCE = 2\gamma$. Considering the sum of the angles in the triangle *ABC*, this implies

$$120° + 2\gamma + \gamma = 180° \Leftrightarrow \gamma = 20°,$$

and therefore $\beta = 2\gamma = 40°$.

Solution for Problem 14.19:

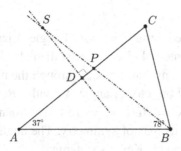

We have

$$\gamma = 180° - (\alpha + \beta) = 180° - (37° + 78°)$$

$$= 65°.$$

Since $\gamma > \alpha$, we also have $AB > BC$. Let *P* denote the common point of the internal bisector of β and *AC*. Every internal bisector in a triangle divides the opposite side in the ratio of the adjacent sides, and we therefore have $AP : PC = AB : BC$. We therefore certainly know that *P* is an internal point of the line segment CD, and that S therefore lies outside ABC. From that we get

$$\angle DSB = \angle DSP = 90° - \angle SPD.$$

Since

$$\angle BPC = 180° - \left(\frac{78°}{2} + 65°\right) = 180° - 104° = 76°,$$

we then have

$$\angle DSB = 90° - 76° = 14°.$$

Solution for Problem 14.20:

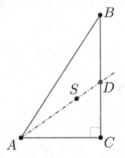

Let us assume that the internal bisector of angle A passes through the centroid S of ABC. The internal bisector must then also be the median of ABC through A and this means that it passes through the midpoint D of the side BC. Since the internal bisector of angle A divides BC in the ratio $AB : AC$, it follows that $AB:AC = BD:DC = 1:1$. The triangle ABC is therefore isosceles with AD as its axis of symmetry. The angle in C must therefore be acute, which contradicts Kevin's statement.

Chapter 15

Quadrilaterals and Polygons

In this chapter, we will take a look at some problems dealing with geometric figures with more than three sides. As usual in geometry, we will be dealing with the fundamental geometric properties of distance, angle and area, but there will also be some problems that have more to do with counting or matters of relative position.

PROBLEMS

Regular Polygons

Problem 15.1:
The sides of a regular pentagon are extended to form a regular five-sided star (a pentagram). The star is composed of the original pentagon and five triangles. Is the area of the pentagon larger than the sum of the areas of the five triangles, equal to the sum of their areas or smaller? Why?

(O-97-6)

Problem 15.2:

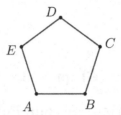

If the vertex E of a regular pentagon $ABCDE$ is reflected on the line joining A and D, we obtain a point P. Determine the angle between AB and BP.

(S5K-98-10)

Problem 15.3:
We are given a square $ABCD$ with sides of length 4 cm. A regular pentagon $ABEFG$ shares a side externally with $ABCD$ as shown, as does an equilateral triangle ABH in the interior of $ABCD$. Determine the angle contained by AE and EH. Why do we know without calculation that the distance between points E and H is certainly less than 8 cm?

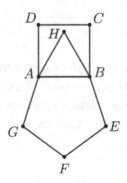

(StU-05-B3)

Problem 15.4:
$ABCDEF$ is a regular hexagon. P and Q are the midpoints of sides AB and EF, respectively. What fraction of the area of the hexagon does the quadrilateral $APQF$ cover?

(O-03-5)

Problem 15.5:

The area of the regular hexagon $ABCDEF$ in the figure is equal to $30\,\text{cm}^2$. B, D and F are the midpoints of PQ, QR and RP, respectively. Determine the area of the equilateral triangle PQR.

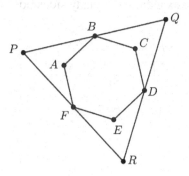

(StU-14-A8)

Problem 15.6:

In a regular hexagon $ABCDEF$, X is the midpoint of the side DE. P and Q are the points in which the diagonal CF meets BX and AX, respectively. What percentage of the area of $ABCDEF$ is the area of the quadrilateral $ABPQ$?

(StU-96-B7)

Problem 15.7:

A superstitious prince owns an estate in the shape of a regular hexagon. Each side of the hexagon is $10\,\text{km}$ long. He decides to retire and give each of his three children part of his estate. Because of his superstition, the piece he gives to each of his children must also be in the shape of a regular hexagon. Also, each of these pieces should be of the same size. Whatever is left will go to the emperor, and he therefore wishes for the hexagonal pieces to be as large as possible. How long is each side of the small hexagons? What fraction of his original estate will go to the emperor?

(O-98-4)

Problem 15.8:

A square *ABCD* and a regular hexagon *BPQRSC* have a common side *BC*. We can fit a regular polygon in the "space" between the square and the hexagon, that has the side *CD* in common with the square and the side *CS* in common with the hexagon. How many sides does this polygon have?

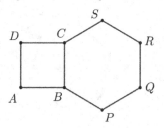

(StU-99-A4)

Problem 15.9:

We are given a regular octagon *ABCDEFGH* with sides of length 1 cm. Determine the area of the triangle *ABF*.

(O-99-4)

Squares

Problem 15.10:

In the figure, we see a square *ABCD* and an equilateral triangle *ABE*. Determine the size of the angle $\angle AED$.

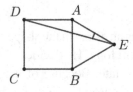

(StU-92-2)

Problem 15.11:

A point *P* is placed in the interior of a square *ABCD* in such a way that *ABP* is an equilateral triangle. Let *X* denote the point of intersection of *AC* and *BP*. How large is the angle contained by the line segments *XA* and *XB*?

(StU-09-A2)

Problem 15.12:

The area of the big square in the figure is equal to $18\,\text{cm}^2$. Each of the dashed lines ends in the corners of the square or the midpoints of its sides. Determine the area of the shaded square.

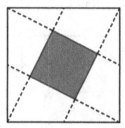

(S5K-00-10)

Problem 15.13:

Determine the ratio of the area of the shaded square to the area of the large square, if we are given that the diagonal lines divide the sides of the large square into four pieces of equal length.

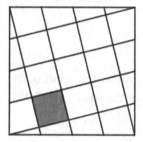

(S5K-01-16)

Problem 15.14:

What percentage of the area of the large square is the taken up by the area of the shaded square in the figure?

(S5K-01-6)

Problem 15.15:

We are given a square $ABCD$. Points P and Q lie on the sides AB and AD in such a way that they are the same distance from A as the midpoint M of the square. How large is the area of the triangle APQ if the area of the square $ABCD$ is equal to $24\,\text{cm}^2$?

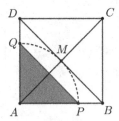

(S5K-16-7)

Problem 15.16:

In a square $ABCD$, P is the midpoint of the side CD. Q is the common point of the lines AP and BD. Determine the ratio of the areas of the triangle BPQ and the square $ABCD$.

(StU-98-B7*)

Problem 15.17:

The points P, Q, R and S are the midpoints of the sides of the square $ABCD$. What fraction of the area of the square $ABCD$ is the gray area shown in the figure?

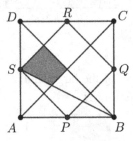

(StU-97-A9)

Problem 15.18:

In the figure, we see a square in which the midpoint of each side is connected to the endpoints of the opposite side by a line segment. These segments enclose an eight-pointed star. What fraction of the area of the square is occupied by the star?

(StU-10-A7)

Problem 15.19:

The square *ABCD* in the figure has sides of length 8 cm. The square is cut up into four triangles as shown. Both the triangle *ABF* and the triangle *AED* have areas measuring 16 cm². What is the area of the triangle *CDE*?

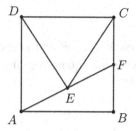

(S5K-07-9)

Problem 15.20:

A figure is composed of three squares *ABGH*, *BCFG* and *CDEF* with sides 8 cm in length. Determine the area of triangle *ADX*.

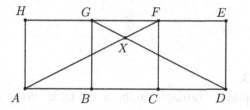

(O-95-3)

Problem 15.21:
A square piece of paper with sides of length 24 cm is folded in such a way that the corner C comes to lie on the midpoint of the side AB, as shown in the figure. The endpoints of the resulting crease are labeled E and F. Determine the length x of the line segment BF.

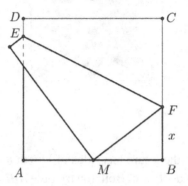

(StU-16-B7)

Problem 15.22:
$ABCD$ is a square with area 36 cm^2. P and Q are equidistant from B. The segments DP and DQ divide the square into three sections with the same area. How far is P from A?

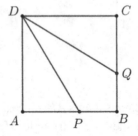

(S5K-09-6)

Rectangles

Problem 15.23:
Pauli Pooli is playing billiards on the 6×4 table shown in the figure. He shoots a (very small) ball from point A on the edge of the table (the "rail" or "cushion"). The ball hits the rail for the first time on the right-hand side as

shown, and it then ricochets off the rail at the same angle as it approached. How often does the ball hit the rail before it returns to A for the first time if it can be hit hard enough? (Your answer should not include the times it touches in A.)

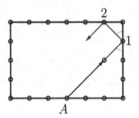

(S5K-11-3)

Problem 15.24:
A rectangle is divided into six squares as shown in the figure. The gray square has an area of $4\,\text{cm}^2$. What is the circumference of the rectangle?

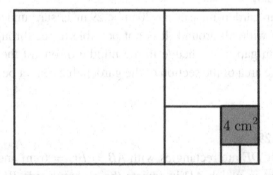

(S5K-11-1)

Problem 15.25:
A rectangle with integer side lengths is cut up into unit squares by lines parallel to its sides. A diagonal of the rectangle is then added to the drawing. How many of the unit squares are cut by the diagonal if the rectangle measures (a) 9×5, (b) 19×95, (c) 19×96?

(StU-95-5)

Problem 15.26:
A rectangle is divided into nine smaller rectangles by horizontal and vertical lines as shown. The areas of some of the small rectangles are given in the figure. Determine the area of the rectangle with the question mark.

	?	6
3		4
1	2	

(StU-01-A5)

Problem 15.27:
We are given a rectangle $PQRS$ with $PQ = 2 \times PS$. Let T be the midpoint of PS and U the midpoint of PQ. Furthermore, let V be the common point of SU and QT. Determine the ratio of the areas of $QRSV$ and PQT.

(O-93-3)

Problem 15.28:
A rectangular garden measuring $20\,m \times 29\,m$ is surrounded by a hedge which is $1\,m$ wide all around. It is not possible to see through the hedge. There is a $3\,m$ gap in the hedge in the middle of one of the longer sides. Determine the area of the section of the garden that cannot be seen from the outside.

(O-96-4)

Problem 15.29:
$ABEF$ and $BCDE$ are rectangles with $AB = BC = 6\,cm$ and $BE = 5\,cm$. X is the point in which AD intersects the common side BE and Y is the common point of AD with the diagonal CE of $BCDE$. How long is the segment XY?

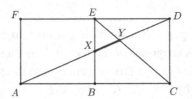

(O-01-5)

Problem 15.30:

In the rectangle $ABCD$, the diagonals are twice as long as the side AB. The point E lies on the side BC in such a way that $AB = BE$ holds. S is the common point of the line AE with the diagonal BD. Determine the angle contained by the lines SA and SB.

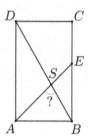

(S5K-08-6)

Problem 15.31:

A rectangle $ABCD$ with the area $12\,\text{cm}^2$ is three times as long as it is wide. The points P and Q lie on the side CD in such a way that $CP = PQ = QD$ holds. S is the common point of the lines AP and BQ. Determine the lengths of the sides of the rectangle $ABCD$ and the area of the triangle ABS.

(StU-97-B3)

Problem 15.32:

In the figure, we have $DP = 3\,\text{cm}$ and $AD = 4\,\text{cm}$, as shown. Determine the area of the rectangle $ABCD$.

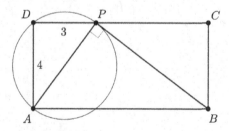

(StU-16-B5)

Other Shapes

Problem 15.33:
We are given the lengths of three of the four sides of a quadrilateral as $a = 35\,\text{mm}$, $b = 60\,\text{mm}$ and $c = 135\,\text{mm}$. Which of the lengths 37 mm, 64 mm, 86 mm and 210 mm is not a possible length of the fourth side? Why is this the case?

(StU-10-A8)

Problem 15.34:
What is the largest number of obtuse angles a quadrilateral can have?

(StU-00-A6)

Problem 15.35:
The rectangle $ABCD$ in the figure is inscribed in the rhombus $EFGH$. Determine the sum of the marked angles.

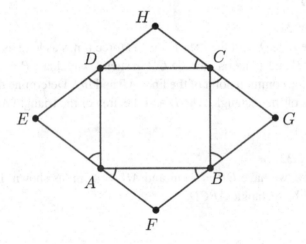

(StU-92-1)

Problem 15.36:

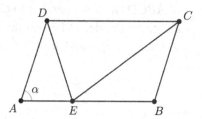

In the parallelogram *ABCD*, the point *E* lies on the side *AB* in such a way that $AD = ED$, $CD = CE$ and $BE = BC$. Determine the angle $\alpha = \angle BAD$.

(StU-99-B4)

Problem 15.37:

Squares *ACDE* and *BCFG* are drawn externally on the sides *AC* and *BC* of the isosceles triangle *ABC*. The segments *AD* and *DF* contain an angle of 98°. Determine the angle $\angle BAC$.

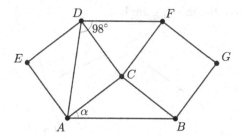

(StU-03-A7)

Problem 15.38:

In a trapezoid *ABCD* with $AB \parallel CD$, we are given $AD = DC = BC$. The point *S* is the common point of the diagonals in the trapezoid, and we are given $\varepsilon = \angle ASB = 135°$. Determine the angle $\alpha = \angle BAD$.

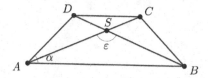

(StU-17-B2)

Problem 15.39:

In the isosceles trapezoid *ABCD* (with *AB* parallel to *CD*), the midpoint *M* of *AB* is the center of the circumcircle. *P* lies on the (shorter) arc *CD* in such a way that *CP* ∥ *MD* holds and the angles ∠*MDC* and ∠*CMP* are equal. Determine their measure.

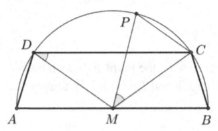

(StU-14-B6)

Problem 15.40:

In the figure, we see a pentagon with the four external angles 68°, 80°, 59° and 85°. Determine the missing angle *x*.

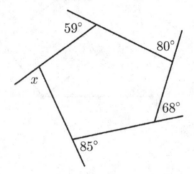

(StU-14-A2)

Problem 15.41:

A point inside the square $ABCD$ is connected to the vertices as shown. The areas of the triangles ABP, BCP, CDP and DAP are named A_1, A_2, A_3 and A_4, respectively. Prove $A_1 + A_3 = A_2 + A_4$.

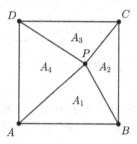

(StU-89-7)

Problem 15.42:

In the figure, we see a quadrilateral $ABCD$ whose diagonals intersect in the point S. The areas of triangles ABS, BCS, CDS and DAS are named A_1, A_2, A_3 and A_4, respectively. Prove $A_1 \times A_3 = A_2 \times A_4$.

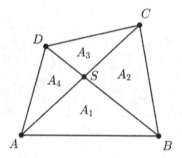

(StU-91-6)

Problem 15.43:

A point P is located in the interior of a parallelogram $ABCD$. Points E, F, G and H are the feet of P on the sides AB, BC, CD and AD, respectively. Prove that the value of $PE + PF + PG + PH$ is independent of the choice of P. Determine the value of this expression for a parallelogram $ABCD$ with $AB = 10\,\text{cm}$, $BC = 6\,\text{cm}$ and $h_a = 5\,\text{cm}$. (Note that h_a is the altitude of the parallelogram perpendicular to the side $a = AB$.)

(StU-15-B3)

Problem 15.44:

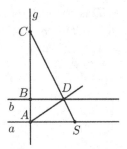

Points A, B and C lie on the line g as shown, and lines a and b are perpendicular to g. We are given $AS = \frac{1}{2}$, $AC = 1$ and $\angle CAD = 45°$. Prove that $AB = \frac{1}{3}$ holds.

(StU-13-B1)

Problem 15.45:

In the two figures, we see squares from which pieces have been cut off, leaving the gray areas. On the left, two equal sections have been cut off, and on the right, four equal sections have been cut off. Which of the gray areas is larger? Why?

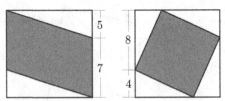

(StU-91-1)

Problem 15.46:

How many times the area of the white part of the figure is the gray part?

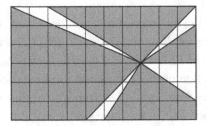

(StU-99-A10*)

Problem 15.47:
A point X lies of the diagonal BD of a parallelogram $ABCD$. The line through X parallel to AD intersects AB in P, and the line through X parallel to AB intersects AD in Q. Prove that the area of the quadrilateral $APCQ$ is exactly half of the area of $ABCD$.

(StU-94-3)

Problem 15.48:
We are given a parallelogram $ABCD$. Points M and N lie on the sides BC and AD respectively, in such a way that $BM = DN$ holds. Let P be an arbitrary point on the side DC. Furthermore, let K be the common point of MN and PA and L the common point of MN and PB. Prove that the area of the triangle PKL is equal to the area of the sum of the areas of triangles ANK and BML.

(StU-93-5)

Problem 15.49:
The following measures are given in the figure: $AB = 20$, $AC = 12$, $\angle ACB = 90°$. D is the midpoint of AB and DE is perpendicular to AB. Determine the area of the quadrilateral $ADEC$.

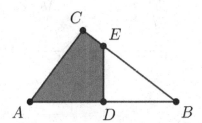

(StU-92-3)

Problem 15.50:

All angles in the figure are right. *M* is equidistant from *A*, *B* and *C*, and the marked distances are of the length 2. Determine the area of the figure.

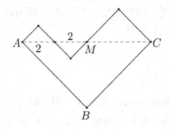

(S5K-09-10)

Problem 15.51:

We are given an isosceles trapezoid *ABCD* with sides of length *a*, *b*, *c* and *d*. (Note that the figure is not drawn to scale.) We know that $b = d$, $a - c = 32$ and $b + d = 42$ hold. The point *X* lies on *AB* in such a way that the circumference of the triangle *AXD* and the circumference of the quadrilateral *BCDX* are equal. How great is the distance from *X* to *B*?

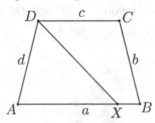

(S5K-15-6)

Problem 15.52:

The corners of the quadrilateral *ABCD* shown in the figure lie on the circle *k* with midpoint *M* and radius $r = 5$ cm. We know that $BC = 6$ cm holds. Determine the length of *AD*.

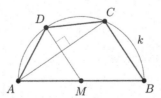

(StU-01-A10)

Problem 15.53:
The quadrilateral in the figure is a trapezoid with $AB \| CD$. Let E and F denote the points in which the line parallel to AD through C intersects AB and BD, respectively. How long is the line segment AB if we are given $BE = 5\,\text{cm}$, $BF = 4\,\text{cm}$ and $FD = 5\,\text{cm}$?

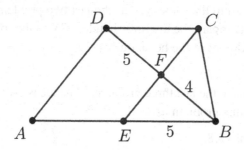

(StU-04-A10A)

TIPS

Tip for Problem 15.1:
Connect each corner of the pentagon to the midpoint of the pentagon. Now the pentagon is also made up of five triangles. Compare these to the ones outside the pentagon.

Tip for Problem 15.2:
First of all, find the measure of the interior angles of the pentagon. It will then be useful to make note of various isosceles triangles.

Tip for Problem 15.3:
The measure of the interior angles of the pentagon will be useful here as well. Also, take a look not just at the obvious triangle AEH but also the triangles ABE and EBH.

Tip for Problem 15.4:
The regular hexagon can be divided into a number of equilateral triangles. Try adding lines parallel to the sides of the hexagon through all relevant points to your drawing and then take a closer look.

Tip for Problem 15.5:
This can be solved in the same way as the last problem. It will be simpler if you try cutting everything into identical triangles of a slightly more sophisticated type, though.

Tip for Problem 15.6:
This can be solved in the same way as the last two problems, but you may find it easier to compare the area of triangle ABX to that of ABM, with M denoting the midpoint of the hexagon.

Tip for Problem 15.7:
Take another look at the last three problems. The ideas used here will prove quite useful for this problem as well.

Tip for Problem 15.8:
You can treat this problem like a word problem with one variable. If a regular polygon has n sides, how big are its internal angles in terms of n?

Tip for Problem 15.9:
Try adding lines perpendicular to AF through G and H to your drawing.

Tip for Problem 15.10:
Note that AED is an isosceles triangle.

Tip for Problem 15.11:
Take a look at the angles in triangle BCX.

Tip for Problem 15.12:
This is easier that it may seem at first glance. There are more small squares like the gray one to be found here, if you look at the figure the right way!

Tip for Problem 15.13:
If you understand the idea used to solve the last problem, this one will prove to be just as simple to solve.

Tip for Problem 15.14:
Note the role that the radius of the circle plays with respect to each of the squares.

Tip for Problem 15.15:
This problem is very closely related to the last one. The same type of observation will lead you straight to the solution.

Tip for Problem 15.16:
Note the similar triangles BQA and DQP.

Tip for Problem 15.17:
There are similar triangles here that play a similar role to those in the last problem. Note that the shaded area is what remains of a small square after removing a sliver of a triangle. There are several triangles similar to this one to be found in the figure.

Tip for Problem 15.18:
The star can be thought of as the part of the surrounding square that remains after removing eight congruent triangles. The area of each of these triangles, as a fraction of the area of the square, can be determined in much the same way we dealt with the last two problems.

Tip for Problem 15.19:
There are similar triangles to be found in this constellation as well. Try adding a few strategic lines perpendicular to the sides of the square.

Tip for Problem 15.20:
If you have been paying attention for the last few problems, it should be easy to find the appropriate similar triangles for this one.

Tip for Problem 15.21:
Note that folding a piece of paper does not change the length of line segments. What happens with CF?

Tip for Problem 15.22:
Try expressing the area of one of the triangles in terms of the length of the segment we wish to determine.

Tip for Problem 15.23:
Drawing the path of the ball with a ruler should give you the answer right away.

Tip for Problem 15.24:
You can figure out the lengths of the sides of each of the squares in the figure, if you start with the smallest and work your way up.

Tip for Problem 15.25:
Try making a sketch of a rectangular grid made up of a small number of squares. The 9×5 case should be easy to draw in its entirety. For the larger

cases, note that it makes a difference if one of the numbers of squares along the edges is a divisor of the other!

Tip for Problem 15.26:
Take a look at small rectangles sharing sides. If you compare their areas, you also know something about the relative areas of pairs of rectangles adjacent to them.

Tip for Problem 15.27:
A nice picture will be useful here. You will need a few more lines than the given ones; try adding some parallels to the lines you already have. This one is not so easy at first, but becomes much simpler if you add the right lines.

Tip for Problem 15.28:
There are two small triangular areas that cannot be seen from outside. Note that each of them is similar to a triangle formed by half of the gap in the hedge.

Tip for Problem 15.29:
Take another look at problem 15.27. You can use the same idea here to great advantage.

Tip for Problem 15.30:
Note that triangles ABD and ABE are both of a very special type. Once you know the angles in these two triangles, there is just one small step left to do.

Tip for Problem 15.31:
The triangles ABS and PQS are similar. That should help you determine their altitudes.

Tip for Problem 15.32:
It seems pretty obvious that the Pythagorean theorem will be useful here, but there are also some similar triangles that will help out.

Tip for Problem 15.33:
The shortest distance between two points is the line segment joining the points. What does this tell you about other possible indirect paths from one point to the other?

Tip for Problem 15.34:
Consider the sum of the angles in a quadrilateral.

Tip for Problem 15.35:
The sum of the angles in the rectangle is 360°.

Tip for Problem 15.36:
There are a lot of isosceles triangles in this figure. Also, if one angle in a parallelogram is α, the others are all either α or $180° - \alpha$.

Tip for Problem 15.37:
Note the symmetry of the figure. What do you know about the quadrilateral *ABFD*?

Tip for Problem 15.38:
Once again, symmetry will help a lot. Also, note all the isosceles triangles in the figure!

Tip for Problem 15.39:
There are also a lot of isosceles triangles in this figure that will help you out. Also, don't forget that parallel lines also result in equal angles.

Tip for Problem 15.40:
If you know the external angle in a vertex of a polygon, you can easily determine the internal angle in that vertex and vice versa.

Tip for Problem 15.41:
Using the sides of the square as bases of the four triangles, add the corresponding altitudes to your figure.

Tip for Problem 15.42:
Adding altitudes to the figure will be useful here as well, but try using one of the diagonals of the quadrilateral as the base of all four triangles.

Tip for Problem 15.43:
As in the last two problems, a figure with lines perpendicular to strategic given lines will prove to be very useful.

Tip for Problem 15.44:
There is an isosceles right triangle in this figure. It will be useful to find it. Also, note that triangles *ASC* and *BDC* are similar.

Tip for Problem 15.45:
It is relatively easy to simply calculate the areas of each of the triangles.

Tip for Problem 15.46:
As in the last problem, it is relatively easy to simply calculate the areas of each of the triangles.

Tip for Problem 15.47:
Add line segment *XC* to your figure. Can you see a pair of triangles with the same area? Maybe another such a pair?

Tip for Problem 15.48:
There are two different ways in which the parallelogram is split in half. If you can find these, you will be on the way to a solution.

Tip for Problem 15.49:
The gray quadrilateral is what is left of the triangle when you remove the small white triangle. Note that these two triangles have a common angle in *B*!

Tip for Problem 15.50:
Try adding some strategic lines to your figure. This shape looks like it was left over after part of a square was chopped off, doesn't it?

Tip for Problem 15.51:
Just write the circumference of each figure as the sum of the lengths of the sides. Setting these expressions equal will allow you to simplify your way to an answer.

Tip for Problem 15.52:
Note that *ABC* is a right triangle. That should get you started. Also, there are some more right triangles in the figure! There is also a very useful pair of similar triangles here you should be able to spot right away.

Tip for Problem 15.53:
Once again, look out for similar triangles.

SOLUTIONS

Solution for Problem 15.1:

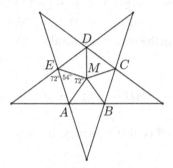

If we connect each of the five corners *ABCD* and *E* of the pentagon with the midpoint *M* of the pentagon, the resulting line segments divide the pentagon into five congruent isosceles triangles. The interior angle of each of these in *M* is equal to $360° \div 5 = 72°$, and the angles at the bases of these isosceles triangles are therefore all equal to

$$(180° - 72°) \div 2 = 54°.$$

Each of the interior angles in the regular pentagon is therefore equal to $54° + 54° = 108°$, and the base angles of each of the triangles drawn outside the pentagon to create the pentagram is therefore equal to $180° - 108° = 72°$. Since $72° > 54°$, the five external triangles therefore have greater altitudes than their interior counterparts, and since they have their bases in common, the external triangles have a larger area than the internal ones. We see that the sum of the areas of the five external triangles is greater than the sum of the areas of the five internal triangles, i.e. the area of the pentagon.

Solution for Problem 15.2:

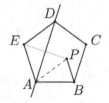

Each interior angle of a regular pentagon is equal to 108°. (This can be calculated by dividing the pentagon into three triangles, the angles in each of which sum to 180°, and then dividing the resulting sum $3 \times 180° = 540°$ by five, for instance.) We see that both the angles $\angle AED$ and $\angle BAE$ are equal to 108°. In the isosceles triangle AED (with $EA = ED$), we therefore have $\angle DAE = \frac{1}{2} \times (180° - \angle AED) = \frac{1}{2} \times (180° - 108°) = 36°$. Because of the symmetry resulting from the reflection, we also have $\angle PAD = \angle DAE = 36°$, and therefore

$$\angle BAP = \angle BAE - \angle PAD - \angle DAE = 108° - 36° - 36° = 36°.$$

Noting that triangle PAB is also isosceles, because of $AB = AE = AP$, we therefore obtain

$$\angle PBA = \frac{1}{2} \times (180° - \angle BAP) = \frac{1}{2} \times (180° - 36°) = 72°.$$

Solution for Problem 15.3:
Since the interior angles of the regular pentagon are all equal to 108°, as we have already seen in the previous problem, and all line segments in the figure are of the same length, namely 4 cm, we can immediately calculate the angles in triangle EBH. Since ABH is an equilateral triangle, we have

$$\angle HBE = \angle HBA + \angle ABE = 60° + 108° = 168°,$$

and therefore $\angle EHB = \frac{1}{2} \times (180° - 168°) = 6°$. This gives us

$$\angle AHE = \angle AHB - \angle EHB = 60° - 6° = 54°.$$

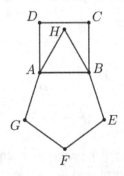

Similarly, since *ABE* is isosceles, we have

$$\angle EAB = \frac{1}{2} \times (180° - \angle ABE) = \frac{1}{2} \times (180° - 108°) = 36°,$$

and therefore

$$\angle EAH = \angle EAB + \angle BAH = 36° + 60° = 96°.$$

In triangle *EAH*, we therefore have

$$\angle HEA = 180° - \angle AHE - \angle EAH = 180° - 54° - 96° = 30°.$$

Finally, the distance between points *E* and *H* is certainly less than 8 cm because of the triangle *EBH*. The sides *BE* and *BH* in this triangle are both 4 cm long. Since the shortest path between two points is always the line segment joining them, the sum of the lengths of two sides is any triangle is always greater than the length of the third side — the Triangle Inequality. In *EBH*, this means $EH < BE + BH = 4\,\text{cm} + 4\,\text{cm} = 8\,\text{cm}$.

Solution for Problem 15.4:

There are many ways to solve this problem, but a relatively easy way is illustrated in the figure. Adding lines parallel to the sides of the hexagon through all vertices of the hexagon and through the midpoints of all the sides creates a triangular grid composed of 24 equilateral triangles in the interior of the hexagon. Since the quadrilateral *APQF* is composed of five of these triangles, *APQF* covers $\frac{5}{24}$ of the hexagon.

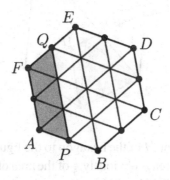

Solution for Problem 15.5:

Once again, we can divide the figure, in this case the triangle PQR, into congruent triangles. Adding the common midpoint M of the triangle and the hexagon, we see that we can divide the entire triangle into 12 congruent triangles, 6 of which make up the hexagon. The area of the triangle is therefore exactly twice that of the hexagon, or $60\,\text{cm}^2$.

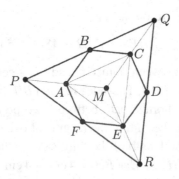

Solution for Problem 15.6:

It is certainly possible to solve this problem in a similar way as we did the last two, but finding appropriate triangles is a bit more difficult here. A different way to solve this problem is as follows.

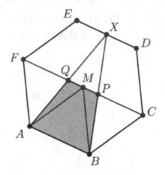

Adding the midpoint M of the hexagon to the figure yields an equilateral triangle ABM, whose area is obviously $\frac{1}{6}$ of the area of the hexagon. Triangle ABX shares the base AB with ABM, and has twice the altitude, as CF divides the hexagon into two identical trapezoids $ABCF$ and $DEFC$. The area of ABX is therefore twice that of ABM, or $2 \times \frac{1}{6} = \frac{1}{3}$ that of the hexagon. We can now note that triangle QPX is similar to ABX, and since Q and P are the midpoints of AX and BX, respectively, the area of QPX is one-quarter that of ABX. This means that the area of the quadrilateral $ABPQ$ is three-quarters of the area of ABX, or $\frac{3}{4} \times \frac{1}{3} = \frac{1}{4}$ the area of the hexagon, i.e. 25% of the area.

Solution for Problem 15.7:
For this problem, we must first decide how best to place the three small hexagons. Since they are to be congruent, it will be best to have them meet with a common vertex in the midpoint of the large hexagon. This is possible due to the fact that the interior angles of the hexagons are all equal to 120°, and $3 \times 120° = 360°$ fills the whole angle in this mid-point. Since we can then divide the hexagons into congruent rhombi as shown, each side of the small hexagons will be half as long as the sides of the large hexagon, or 5 km long. Since three of the rhombi will fall to the emperor while each of the three heirs also receives three such rhombi, $\frac{1}{4}$ of the estate will fall to the emperor.

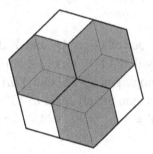

Solution for Problem 15.8:
We first note that the internal angle of the square is equal to 90°, while the internal angle of the regular hexagon is equal to 120°. The angle of the polygon we are searching for in C is therefore equal to $360° - 90° - 120° = 150°$.

Let us assume that the polygon has n sides. Joining all n corners of the polygon to the midpoint M yields n triangles, each of which has an angle sum of 180°. Since these triangles all meet in M, the sum of the angles in the corners of the n-sided polygon is therefore 360° less than the sum of all the interior angles of the triangles, or $180° \times n - 360°$. On the other hand, the sum of all these angles is also equal to $150° \times n$. For the required polygon, we therefore obtain the equation

$$180° \times n - 360° = 150° \times n,$$

which is equivalent to $30° \times n = 360°$, or $n = 12$. The exterior polygon in C therefore has 12 sides; it is a regular dodecagon.

Solution for Problem 15.9:

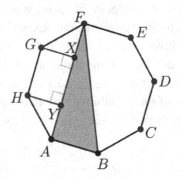

If we let X denote the foot of G on AF and Y denote the foot of H on AF, we see immediately that $XY = GH = 1\,$cm must hold, since $GHYX$ is a rectangle. Furthermore, since XFG and YHA are right isosceles (the internal angles in the octagon are all equal to $135°$, and we therefore have $\angle YAH = \angle BAH - \angle BAY = 135° - 90° = 45°$, for instance), we have

$$FX = YA = \frac{\sqrt{2}}{2} \times 1 = \frac{\sqrt{2}}{2}\,\text{cm}.$$

It therefore follows that

$$AF = FX + XY + YA = \frac{\sqrt{2}}{2} + 1 + \frac{\sqrt{2}}{2} = 1 + \sqrt{2}\,\text{cm}$$

holds, and for the required area of ABF, we therefore obtain

$$\frac{1}{2} \times AB \times AF = \frac{1}{2} \times 1 \times (1 + \sqrt{2}) = \frac{1 + \sqrt{2}}{2}\,\text{cm}^2.$$

Solution for Problem 15.10:

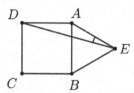

Since the sides of the square $ABCD$ are all equally long, and the same is true of the sides of the equilateral triangle ABE, we have $AD = AB = AE$. This means that triangle ADE is isosceles with $\angle EDA = \angle AED$. Since

$$\angle DAE = \angle DAB + \angle BAE = 90° + 60° = 150°,$$

we therefore obtain

$$\angle AED = \frac{1}{2} \times (180° - \angle DAE) = \frac{1}{2} \times (180° - 150°) = 15°.$$

Solution for Problem 15.11:

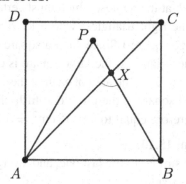

In triangle BCX, we have

$$\angle CBX = \angle CBA - \angle PBA = 90° - 60° = 30°$$

and

$$\angle BCX = \angle BCA = 45°.$$

It therefore follows that

$$\angle BXC = 180° - \angle CBX - \angle BCX = 180° - 30° - 45° = 105°$$

holds, and therefore

$$\angle BXA = 180° - \angle BXC = 180° - 105° = 75°.$$

Solution for Problem 15.12:

Take a look at the following picture:

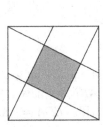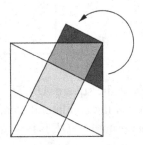

On the right side of the figure, we have rotated the small triangle in the upper right-hand corner by 270°. This results in a position, in which it becomes quite obvious that the small triangle, shown here in dark gray, taken together with the quadrilateral shown in lighter gray, form a square of the same size as the given shaded square. (This can be argued more precisely by taking a closer look at the angles, which must certainly sum to 90°, and the hypotenuses of the right triangles, whose lengths are half the length of the side of the large square.) Since such a square results from rotating each of the four corner triangles, the large square is composed of sections whose areas total to five times the area of each of these smaller squares, and the area of the shaded square is therefore one-fifth of the area of the large square. Its area is therefore equal to $\frac{1}{5} \times 18\,\text{cm}^2 = 3.6\,\text{cm}^2$.

Solution for Problem 15.13:

This problem is quite similar to the last one, and a very similar method will yield the result just as readily.

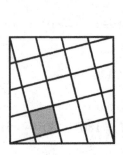

Here, the right side of the figure shows a rotation of the small triangle and the small trapezoid in the upper right by 180°, resulting in a position, in which the four parts at the top of the large square are combined to form two small squares, each the same size as the given shaded square. (This can be argued more precisely, as in the problem above, but the picture should be sufficient to see what is going on.) We therefore have sections whose areas sum to the areas of two small squares on each side of the large square. Together with these $4 \times 2 = 8$ such squares on the edge, there are also nine small squares in the interior of the large square, and the area of the large square is therefore equal to the total area of $8 + 9 = 17$ small squares. The required ratio is therefore equal to $1 : 17$.

Solution for Problem 15.14:

If we assume that the length of the sides of the shaded square is equal to 1, the length of its diagonals is equal to $\sqrt{2}$. If we take a close look at the circular arc, we see that the radius of this arc is equal to both the length of the diagonals of the shaded square and the sides of the larger square.

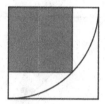

This means that the length of the sides of the large square is then also equal to $\sqrt{2}$. The area of the small shaded square is therefore equal to $1^2 = 1$ and the area of the larger square is equal to $\sqrt{2}^2 = 2$. The area of the shaded square is therefore 50% of the area of the larger square.

Solution for Problem 15.15:

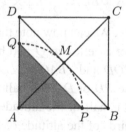

As in the last problem, we can immediately see that the radius of the circular arc plays a role with respect to each of the figures we want to connect. On the one hand, OM is equal to half the length of the diagonal of $ABCD$. On the other hand, it is equal to the lengths of the sides of the right triangle APQ. Since the area of $ABCD$ is given as $24\,\text{cm}^2$, its sides are each $\sqrt{24}\,\text{cm}$ long. The length of the diagonal AC is therefore equal to

$$AB \times \sqrt{2} = \sqrt{24} \times \sqrt{2} = 2 \times \sqrt{12}\,\text{cm}.$$

Since AM is half as long as AC, we see that the length of the radius of the arc is equal to $\sqrt{12}\,\text{cm}$, we also have $AP = AQ = \sqrt{12}\,\text{cm}$, and the area of the triangle APQ is therefore equal to

$$\frac{1}{2} \times (\sqrt{12})^2 = \frac{1}{2} \times 12 = 6\,\text{cm}^2.$$

Solution for Problem 15.16:

In order to solve this problem, we first note that the area of the triangle BPQ results by subtracting the area of BQA from the area of BPA.

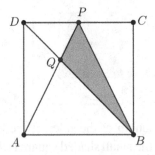

If we assume that the lengths of the sides of $ABCD$ are equal to 1, the area of $ABCD$ is equal to $1^2 = 1$. The area of BPA is equal to

$$\frac{1}{2} \times AB \times BC = \frac{1}{2} \times 1 \times 1 = \frac{1}{2},$$

or half the area of the square. We can now note that triangles BQA and DQP are similar, since they share a vertex in Q and have parallel sides AB and PQ, yielding $\angle AQB = \angle PQD$ and $\angle BAQ = \angle DPQ$ and $\angle QBA = \angle QDP$, respectively. Since the base PD of DQP is half as long as the base AB of BQA, the altitude of DQP is half of the altitude of BQA, and the altitude of BQA is therefore two-thirds of the altitude of BPA, which is equal to the sum of the altitudes of BQA and DQP. Since triangles BQA and BPA share the base AB, the area of BQA is therefore two-thirds of the area of BPA, and the area of BPQ is thus equal to one-third of the area of BPA. It follows that the area of *BPQ* is one-third of one-half of the area of the square, or one-sixth of the area of *ABCD*.

Solution for Problem 15.17:

If we let *M* denote the center of the square *ABCD* and points *X*, *Y* and *Z* as shown, we immediately see that the triangles *XYS* and *MYB* are similar right triangles, as they share an angle in *Y*. Since *MB* is twice as long as *XS*, this means that *XY* is half as long as *YM*, and therefore one-third as long as *XM*. The area of triangle *SXY* is therefore one-sixth of the area of the square *SXMZ*, and the shaded area is therefore five-sixths of the area of this square.

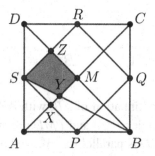

We now note that the area of this square is one-eighth of the area of *ABCD*. This follows from the fact that there are four such squares in the interior of *ABCD*, and two half-squares on each of the four edges of *ABCD*, for a total of eight. The area of the shaded quadrilateral *SYMZ* is therefore five-sixth of one-eighth of the area of *ABCD*, or $\frac{5}{48}$ of the area of the large square.

Solution for Problem 15.18:

The square is made up of the star and eight surrounding congruent triangles. We can determine that area of these triangles by a method quite similar to that used in the last two problems. To this end, let us assume that the sides of the square are of length 1. In the following figure, the small triangle in the upper left has been constructed, along with a second square to the right of the original one.

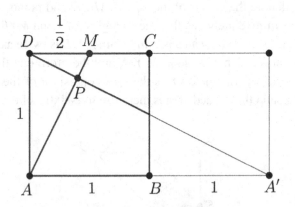

Point A' is on the continuation of AB, with $AB = A'B = 1$. As in the previous problems, triangles PAA' and PMD are similar, as they share the angle in P, and since MD is parallel to AA', we have $\angle APA' = \angle MPD$ and $\angle A'AP = \angle MDP$ and $\angle PA'A = \angle PDM$, respectively. Since the length of the base MD is $\frac{1}{2}$ and the length of the base AA' is 2, and therefore four times as large, the altitude of PMD is also one-quarter the altitude of PAA', or one-fifth of the sum of these altitudes, which is equal to the length of AD, or 1. The area of triangle PDM is therefore equal to

$$\frac{1}{2} \times MD \times \frac{1}{5} = \frac{1}{2} \times \frac{1}{2} \times \frac{1}{5} = \frac{1}{20}.$$

Since there are eight such triangles that are part of the square but not the star, the area of the star is therefore equal to

$$1^2 - 8 \times \frac{1}{20} = \frac{3}{5}$$

of the area of the square.

Solution for Problem 15.19:

Since the area of ABF can be calculated as $16\,\text{cm}^2 = \frac{1}{2} \times AB \times BF$, we obtain $BF = 4\,\text{cm}$, and similarly, since the area of AED can be calculated as $16\,\text{cm}^2 = \frac{1}{2} \times AD \times XE$, we also obtain $XE = 4\,\text{cm}$. Because AYE and ABF are similar right triangles, the fact that $XE = AY$ is half as long as AB implies the fact that EY is half as long as FB, and therefore equal to $2\,\text{cm}$. The altitude of CDE relative to the base CD is therefore equal to $CB - EY = 6\,\text{cm}$, and the area of CDE can therefore be calculated as $\frac{1}{2} \times 8 \times 6 = 24\,\text{cm}^2$.

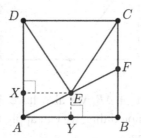

Solution for Problem 15.20:

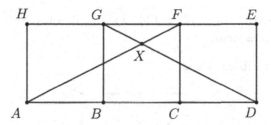

Triangles ADX and FGX are once again similar; we have come across this situation before. They share the angle in X, and since FG is parallel to AD, we have $\angle AXD = \angle FXG$ and $\angle DAX = \angle GFX$ and $\angle XDA = \angle XGF$, respectively. Since AD is three time as long as FG, the altitude of ADX is three times that of FGX, and therefore three-quarters of the total altitude $8\,\text{cm}$ of the figure. The altitude of ADX is therefore equal to $\frac{3}{4} \times 8 = 6\,\text{cm}$, and the area of ADX is therefore equal to

$$\frac{1}{2} \times AD \times 6 = \frac{1}{2} \times 24 \times 6 = 72\,\text{cm}^2.$$

Solution for Problem 15.21:

We first note that MF results by folding CF, and we therefore have

$$MF = CF = BC - x = 24 - x.$$

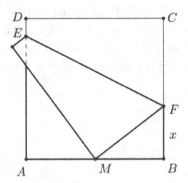

Since M is the midpoint of AB, we also have $MB = \frac{1}{2} \times AB = \frac{1}{2} \times 24$ cm $= 12$ cm. Having noted this, we can apply the Pythagorean theorem in triangle BFM, obtaining

$$BF^2 + BM^2 = MF^2,$$

which is equivalent to $x^2 + 12^2 = (24-x)^2 = 576 - 48x + x^2$, or $48x = 432$, which yields $x = 9$ cm.

Solution for Problem 15.22:

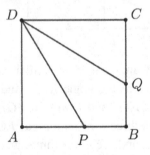

Since $ABCD$ is a square with area 36 cm^2, the sides of the square have the length 6 cm. If we let x denote the length of the segment AP, we see that the area of the triangle APD is equal to $\frac{1}{2} \times AD \times AP = \frac{1}{2} \times 6 \times x = 3x$. Since all three sections have the same area, this is exactly one-third of the area of the square, and we obtain $3x = \frac{36}{3} = 12$, or $x = 4$ cm.

Solution for Problem 15.23:

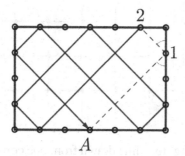

Since the sides of the rectangular table are divided into 4 and 6 sections of equal length, respectively, the total path of the pool ball can easily be drawn, with each section of the path connecting two points on the cushions. Taking a close look at the figure, we see that the ball touches the cushion 9 times before it returns to *A*.

Solution for Problem 15.24:

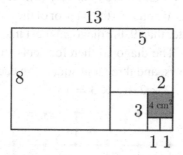

The length of each side of the square with area 4 cm² is obviously 2 cm. This means that the sides of each of the small squares below it are each 1 cm long. From this, we see that the square to the left of the shaded square must have sides of length $2 + 1 = 3$ cm. Continuing in this vein, the square above the shaded square therefore has sides of length $3 + 2 = 5$ cm, and the big left-hand square has sides of length $3 + 5 = 8$ cm. The vertical sides of the rectangle are therefore each 8 cm long, and the horizontal sides are each $8 + 5 = 13$ cm long. The circumference of the rectangle is therefore equal to $13 + 8 + 13 + 8 = 42$ cm.

Solution for Problem 15.25:

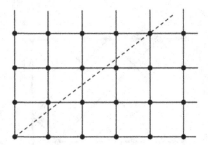

As we can see in the figure, a line drawn from one corner of a grid intersects a certain number of the unit squares in the bottom row (in this case, two) and one more (in this case, three) in each of the following rows. This holds as long as the line does not meet any of the grid vertices. As we can see in the right figure, this causes an interruption in the pattern, and only the smaller number of unit squares are cut in each of the rows adjacent to the grid point in question. We can therefore consider each of the three cases, applying this fact.

(a) Since the numbers 9 and 5 are relatively prime, the diagonal does not pass through any grid points in the interior of the rectangle. Furthermore, since 2×5 is greater than 9, the diagonal must intersect two vertical grid lines in each row. The diagonal therefore cuts two unit squares in the bottom and top rows, and three unit squares in each of the intermediate rows, for a total of $2 \times 2 + 3 \times 3 = 13$.

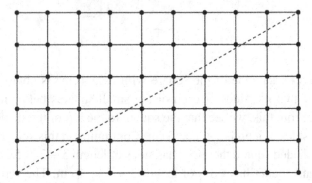

(b) Noting that $95 = 19 \times 5$, we see that the diagonal goes up one square for every five squares it goes to the right. After passing through a grid point, the same thing happens in the next row, and so on. The diagonal therefore cuts 5 unit squares in each of the 19 rows, for a total of $19 \times 5 = 95$.

(c) Since 19 and 96 are relatively prime, the diagonal does not pass through any interior grid points. This means that the diagonal cuts all of the 18 internal horizontal grid lines and all of the 95 internal vertical grid lines in different points, giving $18 + 95 = 113$ points of intersection. Noting that the diagonal in each of these 113 points passes from one cut unit square to the next one, we have a total of 114 unit squares cut by the diagonal.

Solution for Problem 15.26:

	?	6
3		4
1	2	

Since the small rectangle with the area 2 shares a side with the small rectangle with the area 1, it must be twice as wide. This means that the area of the rectangle immediately above it must also be twice as wide as the one immediately above the one with area 1, and since this has the area 3, the area of the middle rectangle is twice 3, or 6. Similarly, as the rectangle with area 4 shares a side with the rectangle above it, and this has an area half as large again, i.e. 6, the rectangle with the question mark must also have an area half as large again as the one immediately below it. We already know that this middle rectangle has the area 6, and the area of the rectangle with the question mark is therefore equal to $6 \times 1.5 = 9$.

Solution for Problem 15.27:

Since $PQ = 2 \times PS$ holds, we can add the rectangle $PQR'S'$ symmetric to $PQRS$ as shown in the figure, and obtain a square $SS'R'R$. Letting PS equal 1, the sides of this square are therefore 2 units in length, and the area of $SR'R$ is half the area of the square, and therefore equal to $\frac{1}{2} \times 2 \times 2 = 2$. We define T' and V' as the points symmetric to T and V with respect to U, respectively. We then see that TQ is parallel to PT'. Since T is the midpoint of SP, V must be the midpoint of SV', and by reasons of symmetry, V' must also be the mid-point of VR'. From this, it follows that the length of VR' is equal to two-thirds of the length of SR'.

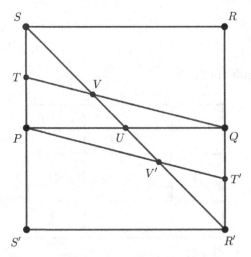

We can now calculate the area of triangle QVR'. Its base QR' is the same length as PS, i.e. one unit. Its altitude is two-thirds of the altitude of RSR', since $R'V$ is two-thirds the length of $R'S$, and therefore equal to $\frac{4}{3}$. The area of QVR' is therefore equal to $\frac{1}{2} \times 1 \times \frac{4}{3} = \frac{2}{3}$, and the area of $QRSV$ is therefore equal to the area of RSR' minus the area of QVR', or $2 - \frac{2}{3} = \frac{4}{3}$. Since the area of PQT is equal to $\frac{1}{2} \times 2 \times \frac{1}{2} = \frac{1}{2}$, the required ratio is equal to $\frac{4}{3} : \frac{1}{2} = 8 : 3$.

Solution for Problem 15.28:

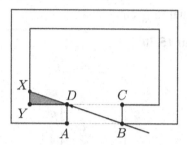

We first note that the gap in the hedge is a rectangle $ABCD$ measuring $1\,\mathrm{m} \times 3\,\mathrm{m}$. Naming the corners of the small triangle on the left that cannot be seen from outside the hedge X and Y, we see that the triangles XYD and DAB are similar, since their hypotenuses lie on a common line, and their sides are parallel. We therefore have $YD : YX = AB : AD = 3 : 1$. Since the gap in the hedge is in the middle, we now that the length of YD is equal to $\frac{1}{2} \times (29 - 3) = 13$. This gives us $YX = \frac{13}{3}$, and the area of XYD is therefore equal to $\frac{1}{2} \times 13 \times \frac{13}{3} = \frac{169}{6}\,\mathrm{m}^2$. Since there is a second such triangle on the right, the total area of the part of the garden that cannot be seen from outside the hedge is equal to $2 \times \frac{169}{6} = \frac{169}{3} = 56\frac{1}{3}\,\mathrm{m}^2$.

Solution for Problem 15.29:

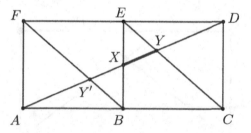

Since X is the midpoint of both BE and CF, the line segment BF is symmetric to EC with respect to X. This line segment intersects AD in a point Y', symmetric to Y with respect to X. Since EC is parallel to FB and FE is of the same length as ED, we see that $Y'Y$ is also of the same length as YD. Similarly, on the other side, it is also of the same length as AY' by analogous reasoning. This means that YY' is one-third the length of the diagonal AD. Since XY is half as long as YY', we see that XY is one-sixth as long as the diagonal AD. It is easy to apply the Pythagorean theorem to calculate the length of AD as $\sqrt{12^2 + 5^2} = \sqrt{144 + 25} = \sqrt{169} = 13$, and the length of XY is therefore equal to $\frac{13}{6}$ cm.

Solution for Problem 15.30:

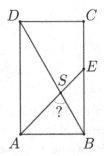

Since $AB = BE$, triangle ABE is a right isosceles triangle, and we have $\angle BAE = 45°$. Also, since BD is twice as long as AB, the right triangle ABD is half of an equilateral triangle, and $\angle DBA = 60°$. From this, consideration of the angles in triangle ABD gives us

$$\angle ASB = 180° - \angle BAE - \angle DBA = 180° - 45° - 60° = 75°.$$

Solution for Problem 15.31:

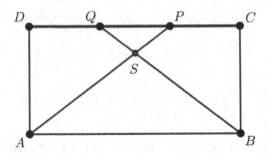

If we let x denote the length of AD, we have $AB = 3x$, and the area of $ABCD$ is therefore equal to $x \times 3x = 3x^2 = 12$, which gives us $AD = x = 2$ cm, $AB = 3x = 6$ cm. Isosceles triangles ABS and PQS are similar, as their sides are pairwise parallel, and their bases are in the ratio $3:1$. The altitude of ABS is therefore three times as large as the altitude of PQS, or one-quarter of AD. The altitude of ABS is therefore equal to three-quarters of 2 cm, or 1.5 cm. The area of *ABS* is therefore equal to

$$\frac{1}{2} \times AB \times 1.5 = \frac{1}{2} \times 6 \times \frac{3}{2} = \frac{9}{2} = 4\frac{1}{2} \text{ cm}^2.$$

Solution for Problem 15.32:

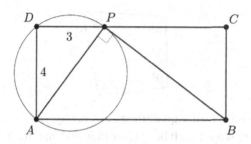

Since *ABCD* is a rectangle, triangle *APD* has a right angle in *D*. By the Pythagorean theorem, we therefore have

$$AP = \sqrt{DP^2 + AD^2} = \sqrt{3^2 + 4^2} = 5.$$

We now note that triangles *APD* and *BAP* are similar, since they are both right triangles and $\angle BAP = 90° - \angle PAD = \angle DPA$ holds. We therefore have

$$AB:AP = AP:DP \Leftrightarrow AB:5 = 5:3,$$

or $AB = \frac{25}{3}$. The area of *ABCD* can now be calculated as

$$AB \times AD = \frac{25}{3} \times 4 = \frac{100}{3} \text{ cm}^2.$$

Solution for Problem 15.33:

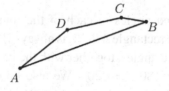

Let us take a look at the quadrilateral *ABCD* shown in the figure. The line segment *AB* is certainly the shortest path connecting *A* to *B*. This means that the sum of the lengths of the sides *AD*, *DC* and *CB* in the quadrilateral must be longer that the length of *AB*, since the polyline *ADCB* is also a path from A to B. This means that the sum of the lengths of any three sides of a quadrilateral must be greater than the length of the fourth. For this reason, the fourth side of a quadrilateral with sides of length 35 mm, 60 mm and 135 mm cannot be 37 mm, as $35 + 37 + 60 = 132 < 135$.

Solution for Problem 15.34:

The sum of the angles in a quadrilateral is always $360°$. If all four angles were obtuse, they would each be greater than $90°$, and the sum of the angles would be greater than $4 \times 90° = 360°$, which yields a contradiction. It is possible for three angles to be obtuse, as we see in the figure. The answer is therefore three.

Solution for Problem 15.35:

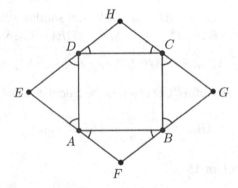

The sum of the marked angles in each of the four points, together with the interior angle of the rectangle $ABCD$, is always $180°$. This means that the total of all these marked angles, together with the interior angles of the rectangle, is equal to $4 \times 180° = 720°$. We also know that the sum of the four (right) angles in the rectangle is equal to $360°$. The sum of the marked angles is therefore equal to $720° - 360° = 360°$.

Solution for Problem 15.36:

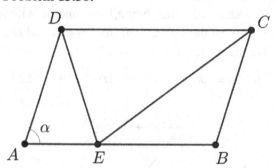

Since AED is an isosceles triangle and we are given $\angle BAD = \alpha$, we have $\angle EAD = \alpha = \angle DEA$, and $\angle ADE = 180° - 2\alpha$. Since $ABCD$ is a parallelogram, we have

$$\angle ADC = 180° - \angle BAD = 180° - \alpha.$$

From this, we obtain

$$\angle CED = \angle EDC = \angle ADC - \angle ADE = (180° - \alpha) - (180° - 2\alpha) = \alpha$$

in the isosceles triangle CDE. Since E lies on the line AB, we now have

$$\angle BEC = 180° - \angle CED - \angle DEA = 180° - \alpha - \alpha = 180° - 2\alpha,$$

and since triangle BEC is also isosceles, we obtain

$$\angle CBE = 180° - 2 \cdot \angle BEC = 180° - 2 \cdot (180° - 2\alpha) = 4\alpha - 180°.$$

On the other hand, since $ABCD$ is a parallelogram, we also have

$$\angle CBE = 180° - \angle BAD = 180° - \alpha.$$

We therefore obtain

$$4\alpha - 180° = 180° - \alpha,$$

which is equivalent to $5\alpha = 360°$, or $\alpha = 72°$.

Solution for Problem 15.37:
Since ABC is isosceles, the entire figure is symmetric with respect to the line perpendicular to AB through C. This means that $ABFD$ is an isosceles trapezoid, and we therefore have

$$\angle BAD = 180° - \angle ADF = 180° - 98° = 82°.$$

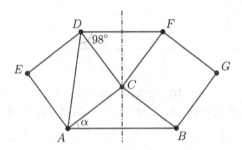

Since AD is the diagonal of the square $ACDE$, we have $\angle CAD = 45°$, and therefore

$$\alpha = \angle BAD - \angle CAD = 82° - 45° = 37°.$$

Solution for Problem 15.38:

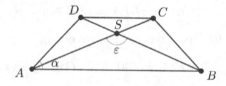

Because of the given identities, the trapezoid is isosceles, as are the triangles SAB, DAC and SCD. Since $\angle CSD = \angle ASB = \varepsilon$, we therefore have

$$\angle BAS = \frac{1}{2}(180° - \varepsilon) = \angle DCS = \angle DAC.$$

We therefore obtain

$$\alpha = \angle BAS + \angle SAD = 180° - \varepsilon = 180° - 135° = 45°.$$

Solution for Problem 15.39:

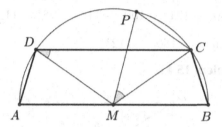

Let us name $\angle MDC = \angle CMP = \alpha$. Since the triangle MCD has two sides that are radii of the circumcircle, we have $\angle MDC = \angle DCM = \alpha$. Furthermore, since CP is parallel to MD, we also have $\angle MDC = \angle PCD = \alpha$, and therefore $\angle PCM = \angle PCD + \angle DCM = 2\alpha$. We now note that the triangle MCP also has two sides that are radii of the circumcircle, and we therefore have $\angle MPC = \angle PCM = 2\alpha$. Since the angles in MCP sum to $180°$, we therefore obtain

$$\angle MPC + \angle PCM + \angle CMP = 2\alpha + 2\alpha + \alpha = 5\alpha = 180°,$$

and therefore $\alpha = 36°$.

Solution for Problem 15.40:

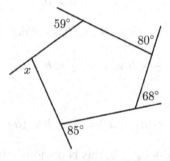

Each of the given external angles is the supplementary angle to the neighboring internal angle. These are therefore equal to $180° - 59° = 121°$, $180° - 80° = 100°$, $180° - 68° = 112°$ and $180° - 85° = 95°$, respectively. Since a pentagon can be divided into three triangles by two diagonals, the sum of the internal angles in any pentagon is always equal to $3 \times 180° = 540°$. The internal angle next to the missing angle x is therefore equal to

$$540° - 121° - 100° - 112° - 95° = 112°.$$

From this we obtain $x = 180° - 112° = 68°$.

Solution for Problem 15.41:

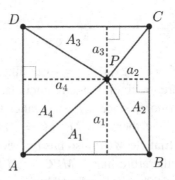

If we add the altitudes a_1, a_2, a_3 and a_4 of the triangles to the figure as shown, we can write the areas of the triangles (each of which has the same base b) as

$$A_1 = \frac{1}{2} \times b \times a_1, \quad A_2 = \frac{1}{2} \times b \times a_2, \quad A_3 = \frac{1}{2} \times b \times a_3 \text{ and}$$

$$A_4 = \frac{1}{2} \times b \times a_4.$$

We wish to prove $A_1 + A_3 = A_2 + A_4$. This is now equivalent to

$$\frac{1}{2} \times b \times a_1 + \frac{1}{2} \times b \times a_3 = \frac{1}{2} \times b \times a_2 + \frac{1}{2} \times b \times a_4,$$

or

$$\frac{1}{2} \times b \times (a_1 + a_3) = \frac{1}{2} \times b \times (a_2 + a_4),$$

and since $a_1 + a_3 = a_2 + a_4 = b$, this is obviously true.

Solution for Problem 15.42:

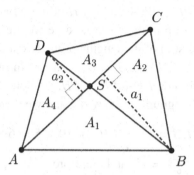

If we add the altitudes a_1 and a_2 of the triangles to the figure as shown, we can write the areas of the triangles as

$$A_1 = \frac{1}{2} \times AS \times a_1, \quad A_2 = \frac{1}{2} \times CS \times a_1, \quad A_3 = \frac{1}{2} \times CS \times a_2 \quad \text{and}$$

$$A_4 = \frac{1}{2} \times AS \times a_2.$$

We wish to prove $A_1 \times A_3 = A_2 \times A_4$. This is now equivalent to

$$\left(\frac{1}{2} \times AS \times a_1\right) \times \left(\frac{1}{2} \times CS \times a_2\right) = \left(\frac{1}{2} \times CS \times a_1\right)$$
$$\times \left(\frac{1}{2} \times AS \times a_2\right),$$

and this is obviously true.

Solution for Problem 15.43:

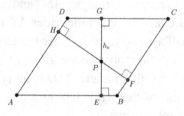

As we can see in the figure, $PE + PG = h_a$ is equal to the length of the altitude of the parallelogram perpendicular to AB and CD. The length of this altitude is obviously independent of the choice of P. Similarly, $PF + PH$ is equal to the length of the altitude of the parallelogram perpendicular to BC

and *DA* and is similarly independent of the choice of *P*. This is therefore also true of the sum $PE + PF + PG + PH$ of these two expressions.

The value of this sum in the concrete case we are asked to calculate is therefore equal to the sum of the lengths of the two altitudes, one of which we are given as $PE + PG = EG = h_a = 5\,\text{cm}$. In order to calculate the length of the other altitude $PF + PH = FH$, we note that we can calculate the area of the parallelogram in two ways, and we therefore have

$$AB \times EG = BC \times FH, \text{ or } 10 \times 5 = 6 \times FH.$$

This gives us $FH = \frac{10 \times 5}{6} = \frac{25}{3}$, and therefore

$$PE + PF + PG + PH = EG + FH = 5 + \frac{25}{3} = 13\frac{1}{3}\,\text{cm.}$$

Solution for Problem 15.44:

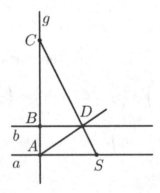

First of all, it is useful to note that triangles *ASC* and *BDC* are similar, since they share an angle in *C* and have parallel sides *AS* and *BD*. Since we are given that $AS = \frac{1}{2}$ is half as long as $AC = 1$, we therefore also know that *BD* is half as long as *BC*. Furthermore, we are given $\angle CAD = 45°$, and since *BD* is perpendicular to *BA*, the triangle *BAD* is an isosceles right triangle with $BD = BA$. This means that *AB* is half as long as *BC*, or one-third as long as *AC*, and we therefore have

$$AB = \frac{1}{3} \times AC = \frac{1}{3} \times 1 = \frac{1}{3},$$

as claimed.

Solution for Problem 15.45:

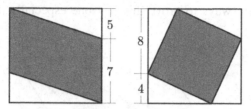

The lengths of the sides of the squares are equal to

$$5 + 7 = 4 + 8 = 12.$$

The area of each of the white triangles in the left figure is therefore equal to $\frac{1}{2} \times 12 \times 5 = 30$, and the gray area is therefore equal to

$$12^2 - 2 \times 30 = 84.$$

The area of each of the white triangles in the right figure is equal to $\frac{1}{2} \times 8 \times 4 = 16$, and the gray area is therefore equal to

$$12^2 - 4 \times 16 = 64,$$

and we see that the gray area on the left is the larger of the two.

Solution for Problem 15.46:

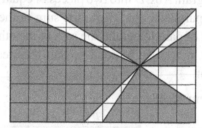

Counting the little squares, we see that the area of the rectangle is equal to $10 \times 6 = 60$. Again counting squares, we see that the area of the triangle in the upper left corner is equal to $\frac{1}{2} \times 2 \times 3 = 3$. Similarly, the areas of the other three triangles (taken clockwise) are equal to $\frac{1}{2} \times 1 \times 3 = \frac{3}{2}$, $\frac{1}{2} \times 2 \times 3 = 3$ and $\frac{1}{2} \times 1 \times 3 = \frac{3}{2}$, respectively. The sum of these white areas is therefore equal to

$$3 + \frac{3}{2} + 3 + \frac{3}{2} = 9,$$

and the sum of the gray areas is equal to $60 - 9 = 51$. The gray part is therefore $\frac{51}{9}$ times as large as the white part.

Solution for Problem 15.47:

There are several ways to prove this, but one especially nice way is the following.

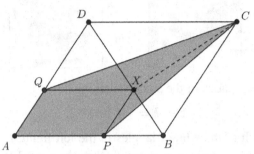

Comparing triangles *BCP* and *BCX*, we see that both share the side *BC*. Their areas must therefore be equal, since *XP* is parallel to *BC*, and the distances of *P* and *X* from the common base *BC*, i.e. their altitudes relative to this base, are equal. Similarly, the areas of *CDQ* and *CDX* are also equal by the same reasoning relative to their common side *CD*. We see that the sum of the areas of *BCP* and *CDQ* is equal to the sum of the areas of *BCX* and *CDX*, and this is the area of the triangle *BCD*. Since the diagonal BD divides the parallelogram *ABCD* into two pieces of equal area, this is exactly half the area of the parallelogram. Since the sum of the areas of *BCP* and *CDQ* is half the area of parallelogram, the area of the other part, namely the quadrilateral *APCQ*, must also be equal to half the area of parallelogram, completing the proof.

Solution for Problem 15.48:

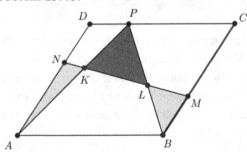

Since we are given that $BM = DN$ holds, *N* and *M* are symmetric with respect to the midpoint of the parallelogram. This means that MN divides the parallelogram into two sections *DNMC* and *BMNA* of equal area, and the area of *BMNA* is therefore equal to half the area of the parallelogram. The area of triangle *ABP* is also equal to half the area of the parallelogram, as they share the base *AB* and the altitude perpendicular to *AB* through *P*. If we write the area of a figure *XYZ* as [*XYZ*], we therefore have

$$\frac{1}{2} \times [ABCD] = [BMNA] = [ABP],$$

and this gives us

$$[BML] + [BLKA] + [KNA] = [BLKA] + [KLP],$$

and therefore

$$[BML] + [KNA] = [KLP],$$

as claimed.

Solution for Problem 15.49:

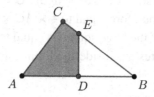

First of all, since *ABC* is a right triangle, the Pythagorean theorem gives us

$$BC = \sqrt{AB^2 - AC^2} = \sqrt{20^2 - 12^2} = 16.$$

We now note that *DBE* is also a right triangle, and that it shares the angle in *B* with *ABC*. The two triangles are therefore similar. We are given $DB = \frac{1}{2} \times AB = \frac{1}{2} \times 20 = 10$, and because of the similar triangles, we also have

$$ED : DB = AC : CB \Leftrightarrow ED : 10 = 12 : 16 \Leftrightarrow ED = \frac{15}{2}.$$

The area of triangle *DBE* is therefore equal to $\frac{1}{2} \times DB \times ED = \frac{1}{2} \times 10 \times \frac{15}{2} = \frac{75}{2}$, and since the area of triangle *ABC* is equal to $\frac{1}{2} \times AC \times BC = \frac{1}{2} \times 12 \times 16 = 96$, the required area of *ADEC* is equal to $96 - \frac{75}{2} = \frac{117}{2}$.

Solution for Problem 15.50:

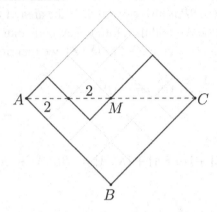

Since all angles in the figure are right and M is equidistant from A, B and C, we can add small squares outside the figure as shown, and the figure can be extended to a large square with sides AB and BC. Also, we can divide the figure itself into small squares, with all of the small squares of the same size. The diagonal of the large square has the length $AC = 2 \times (2+2) = 8$, and the area of triangle ABC is therefore equal to $\frac{1}{2} \times AC \times MB = \frac{1}{2} \times 8 \times 4 = 16$. It follows that the area of the large square is equal to $2 \times 16 = 32$, and since 10 of the 16 small squares are inside the figure, its area is therefore equal to $\frac{10}{16} \times 32 = 20$.

Solution for Problem 15.51:

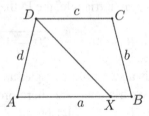

Since the circumference of the triangle AXD is equal to the circumference of the quadrilateral $BCDX$, we know that

$$DA + AX + XD = XB + BC + CD + DX$$

holds. This can be written as

$$d + (a - XB) + XD = XB + b + c + XD,$$

and since $b = d$, this is equivalent to

$$a - XB = XB + c, \text{ or } a - c = 2 \times XB.$$

Since $a - c = 32$ is given, we obtain $XB = 16$. (Note that we did not even use the information about the sum of the lengths of b and d we were given!)

Solution for Problem 15.52:

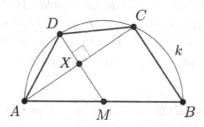

Since C lies on the semi-circle with diameter AB, the angle subtended by the diameter AB from C is right. Triangle ABC is therefore a right triangle, and since we are given $AB = MA + MB = 5 + 5 = 10$ cm and $BC = 6$ cm, we obtain

$$AC = \sqrt{AB^2 - BC^2} = \sqrt{10^2 - 6^2} = 8 \text{ cm}.$$

Since the radius MD of k is perpendicular to the chord AC, it intersects AC in its midpoint X, and we have $AX = 4$ cm. Furthermore, since both MX and BC are perpendicular to AC, triangles AMX and ABC are similar, and we obtain

$$MX : AM = BC : AB \Leftrightarrow MX : 5 = 6 : 10 \Leftrightarrow MX = 3.$$

It follows that $DX = DM - XM = 5 - 3 = 2$ cm holds. Since AXD is a right triangle, the Pythagorean theorem then gives us

$$AD = \sqrt{AX^2 + XD^2} = \sqrt{4^2 + 2^2} = \sqrt{20} \text{ cm}.$$

Solution for Problem 15.53:

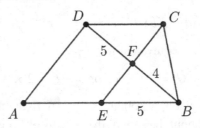

Since we are given both $AB \parallel CD$ and $AD \parallel EC$, we know that $AECD$ is a parallelogram, and that $AE = CD$ therefore holds. In order to calculate the length of CD, we note that triangles FEB and FCD are similar. This follows from the fact that their angles in F are vertex angles and therefore equal, and $\angle BEF = \angle DCF$ are equal because EB is parallel to CD. We therefore obtain

$$CD:DF = EB:BF \Leftrightarrow CD:5 = 5:4 \Leftrightarrow CD = \frac{25}{4}\, \text{cm}.$$

All that remains is to note

$$AB = AE + EB = DC + EB = \frac{25}{4} + 5 = \frac{45}{4} = 11\frac{1}{4}\, \text{cm}.$$

Chapter 16

Divisibility

There are a lot of interesting things we can do with numbers that just result from ordinary division. Here are some enjoyable problems of this type. It will prove to be really useful in solving many of these problems if you are familiar with the rules for deciding under which circumstances a number is divisible by small numbers like 2, 3, 4, 5, 8 or 9.

PROBLEMS

Problem 16.1:
We are given the number 2007 and want to add a two-digit number (≥ 10) to its end. The result should be a six-digit number 2007__, which is divisible by 3, 4 and 5. Which two-digit number do we have to add on?

(S5K-07-5)

Problem 16.2:
Consider two-digit numbers, the sum of whose digits is equal to 9. How many such numbers are divisible by 4?

(StU-05-A1)

Problem 16.3:
How many three-digit numbers exist, that are divisible by 5, but are written without using the digit 5?

(StU-02-B2)

Problem 16.4:
The sum of the digits of a three-digit number abc with $b = c$ is divisible by 7. Does this imply that the number abc must then also certainly be divisible by 7?

(StU-94-1)

Problem 16.5:
We add 266 to the three-digit number $2x3$ (with x being the tens-digit). The result is a three-digit number $5y9$ (with y being the tens-digit of the sum). This resulting number is divisible by 9. Determine the value of $x + y$.

(S5K-99-1)

Problem 16.6:
Determine the smallest four-digit number with an odd ones-digit, which is divisible by 9.

(S5K-07-1)

Problem 16.7:
In a seven-digit-number N divisible by 6, the first, third, fifth and seventh digits are the same, as are the second, fourth and sixth digits (in other words, the decimal representation of N is $abababa$). If we delete the first and the last digit, we obtain a five-digit number (with the decimal representation $babab$), that is divisible by 9, but not by 18. Determine the second digit b of the original seven-digit number.

(StU-08-A10)

Problem 16.8:
Among all five-digit numbers, some like 14589 have rising digits — reading the digits from left to right, they become larger and larger. Determine the smallest five-digit number with rising digits that is divisible by 12.

(S5K-12-5)

Problem 16.9:
Using the digits 2, 0, 1 and 7, we want to write the smallest possible number that is divisible by 36. Each of the digits must be used at least once, and 0 cannot be the lead digit. Determine the number.

(S5K-17-10)

Problem 16.10:
How many digits does the smallest positive integer have, that is divisible by 36 and is written using only the digits 5 and 8?

(StU-15-A6)

Problem 16.11:
The digits from 1 to 9 are written on a blackboard. Peter and Max are playing the following game. Together, they write a nine-digit number, using these digits. Each digit may only be used once. They alternate in choosing a digit among those still available, and placing it in one of the nine possible slots, with Peter going first. If the nine-digit number they end up with at the end is divisible by 36, Max wins. If it is not, Peter wins. In Peter's first move, he chooses the 4, and puts it in the tens-digit slot, i.e. the second from the right. Is there a way for Max to guarantee a win? If so how? If not, why not?

(StU-18-A8)

Problem 16.12:
Which of the numbers 125, 165, 181, 230 and 260 yields a product divisible by 45 when it is multiplied by 474?

(StU-02-A6)

Problem 16.13:
Dekai writes numbers that alternate the digits 5 and 0, like 505 or 50505050. How many digits does the smallest such number have that leaves the remainder 0 when it is divided by 45?

(S5K-15-7)

Problem 16.14:
Which digits can we put in place of the stars in the five-digit number 7∗42∗ if we want the number to be divisible by 45?

(StU-95-2)

Problem 16.15:
We are given a five-digit number *378* in which we do not know the first and last digits. We do, however, know that the number is divisible by 72. Determine all possible numbers with this property.

(StU-90-2)

Problem 16.16:
In a five-digit number, the three-digit number x formed by the first three digits is twice as large as the two-digit number y formed by the last two. What are the possible values of y? What is the first digit of x? Determine all five-digit numbers with this property that are divisible by 81.

(StU-03-B6)

Problem 16.17:
The five-digit number $n = 3 * * * 0$ is divisible by 2004. What is the sum of the digits of n?

(StU-04-A8)

Problem 16.18:
The first digit of a 1998-digit number is 1. Every other digit, taken together with its predecessor, determines a two-digit number, which is divisible by either 19 or 21. Determine the ones-digit of the number.

(StU-98-A2)

Problem 16.19:
The first digit of a 2002-digit number N is 2. Any two successive digits of N form a two-digit number that is divisible by 13. Determine the sum of the digits of N.

(StU-02-B3)

Problem 16.20:
What is the remainder when $2012^2 - 2011 \times 2013$ is divided by 11?

(StU-11-A1)

Problem 16.21:
Is there a number that yields a remainder of 1 after division by 3, a remainder of 2 after division by 4, a remainder of 3 after division by 5 and a remainder of 4 after division by 6?

(StU-90-5)

Problem 16.22:
We are given four positive integers a, b, c and d. Each of these numbers is divided by 5 and the remainders of all four calculations are added. It turns

out that the resulting sum of the remainders equals 3. Prove that at least one of the numbers a, b, c and d must be divisible by 5. Give an example of three different positive integers a, b and c, none of which is divisible by 5, with the property that the sum of the remainders resulting by division of all three by 5 is equal to 3.

(StU-16-B4)

Problem 16.23:
A prime between 10 and 20 is divided by a prime. The result is prime and the remainder is also prime. Determine all possible divisions with this property.

(O-98-3)

Problem 16.24:
A prime number, a perfect square and a third number greater than the perfect square are all smaller than 20. The product of the three numbers is not divisible by any prime greater than three and the sum of the three numbers is a prime greater than 30. Determine the numbers.

(O-97-5)

Problem 16.25:
Two numbers are called "in-laws" if each is divisible by the sum of the digits of the other one. How many two-digit numbers are in-laws of 24?

(O-01-3)

Problem 16.26:
The sum of three consecutive odd positive integers is smaller than 30 and divisible by 9. Determine all possible triples of integers satisfying these conditions.

(StU-13-B2)

Problem 16.27:
Three different positive integers form a "club" if each of them divides the sum of the other two. (An example of such a club is (2, 4, 6), since $2 \mid 10 = 4 + 6$, $4 \mid 8 = 2 + 6$ and $6 \mid 6 = 2 + 4$ hold.) Determine all numbers that form a club together with 8 and 24.

(O-02-6)

Problem 16.28:
Determine two numbers with the product one million, neither of which is written using the digit 0.

(S5K-00-9)

Problem 16.29:
Which of the integers from 1900 to 1999 are not primes and have no divisors (other than 1) less than 30?

(S5K-99-5)

Problem 16.30:
Using the five digits 1, 2, 3, 4 and 5, we wish to create a five-digit code *abcde* with the following properties: The three-digit number *abc* made up of the three front digits is divisible by 4, the three-digit number *bcd* made up of the middle digits is divisible by 5, and the three-digit number *cde* made up of the end digits is divisible by 3. Determine the code.

(S5K-05-10)

Problem 16.31:
The number 312 has three digits, does not include the digit 0, and is divisible by the number 12, which we obtain by erasing the hundreds-digit ($312 \div 12 = 36$). Determine the largest three-digit number with this property.

(S5K-06-10)

Problem 16.32:
How many positive integers smaller than 1000 are there, that are neither divisible by 5 nor by 7?

(StU-93-4)

Problem 16.33:
The sum of 49 (not necessarily different) positive integers is 999. Prove that the greatest common divisor of these numbers cannot be greater than 9.

(StU-94-4)

Problem 16.34:
We consider the product of all positive integers from 19 to 99, i.e.

$$19 \times 20 \times 21 \times \cdots \times 98 \times 99.$$

We now eliminate all the even numbers and all numbers divisible by 5 from this term. Determine the ones-digit of the resulting product.

(StU-99-B2)

Problem 16.35:
We are given the following information about two positive integers a and b. The result of the division $a \div b$, cut off after the second decimal, is 87.43. Furthermore, the integer division yields a remainder of 10. Determine a and b.

(StU-11-B7)

Problem 16.36:
All integers from 1 to 100 are written on a blackboard. In a first step, all numbers divisible by 2 are deleted. Next, all numbers divisible by 3 are deleted. Finally, of the remaining numbers, all divisible by 11 are deleted. How many numbers are left on the board?

(StU-12-A4)

Problem 16.37:
We say that a number k can be "cut away" from a number N divisible by k by subtracting k from N. For example, starting with 35, we can successively cut away the numbers 7, 4 and 6, since 35 is divisible by 7, $35 - 7 = 28$ is divisible by 4, and $28 - 4 = 24$ is divisible by 6. Why can no positive integer exist, from which 5, 4, 3, 2, 1 can be successively cut away in this order? Determine the smallest positive integer, from which 1, 2, 3, 4, 5 can be successively cut away in some order.

(StU-12-B7)

TIPS

Tip for Problem 16.1:
If a number is divisible by 3, 4 and 5, we know a lot about its digits from the laws of divisibility.

Tip for Problem 16.2:
The number is certainly even. You can easily check all possible combinations of digits with sum 9, whose ones-digits are even.

Tip for Problem 16.3:
What must the ones-digit of a number be, if the number is divisible by 5, but the digit 5 is not one of its digits?

Tip for Problem 16.4:
It is not too difficult to write down all numbers of this type and check. On the other hand, you might try to apply some algebra for a more elegant approach!

Tip for Problem 16.5:
Since the sum of the digits of a number divisible by 9 must also be divisible by 9, you already know the value of y.

Tip for Problem 16.6:
Note that a small number may have a large ones-digit. In fact if the sum of the digits is fixed, that will make the other digits, and therefore the number as a whole, smaller!

Tip for Problem 16.7:
If a number is divisible by 6, it is divisible by both 2 and 3. If a number is divisible by 9, but not by 18, the sum of its digits must be divisible by 9, but its last digit must be odd.

Tip for Problem 16.8:
A number is divisible by 12 if it is divisible by both 3 and 4.

Tip for Problem 16.9:
Since $36 = 4 \times 9$, the number we are looking for must be divisible by 4 and 9. If you think about it, you will see that you should care about the 9 first.

Tip for Problem 16.10:
As in the preceding problem, $36 = 4 \times 9$ is again an important factor. What does divisibility by 4 mean? You can work out the answer by proceeding backwards from there.

Tip for Problem 16.11:
This is another problem with 4×9 playing an important role. Once again, if you think about it a bit, you will notice that the 9 doesn't really matter at all.

Tip for Problem 16.12:
In this case, we note $45 = 5 \times 9$.

Tip for Problem 16.13:
Once again, we note $45 = 5 \times 9$. The factor 5 will not matter in this case.

Tip for Problem 16.14:
Another $45 = 5 \times 9$ problem. Check out the options for divisibility by 5 first.

Tip for Problem 16.15:
This is very similar to the last one, except this time we have $72 = 8 \times 9$. Check out the options for 8 first in this case.

Tip for Problem 16.16:
What do three-digit numbers that are twice as big as two-digit numbers look like? Note that 2×40 has only two digits, while 2×60 has three, for instance.

Tip for Problem 16.17:
Note that the multiple of 2004 we are searching for has five digits and begins with the digit 3. That tells us something about our options for multiplying 2004 to get such a number n. Also, the last digit has to be a zero!

Tip for Problem 16.18:
We know the first digit, and the condition gives us only one option for the second, as you can easily figure out. If you think about it some more, each digit forces the value if the next digit in a unique way. Keep at it, and a pattern will emerge.

Tip for Problem 16.19:
Take another look at the last problem. There is a similar pattern in the situation described here.

Tip for Problem 16.20:
Take a closer look at the numbers involved here. What is one more than 2012 and what is one less than 2012? Applying some of your buried algebra might

be useful here, if you don't want to actually calculate everything the long
way.

Tip for Problem 16.21:
Note that 1 is 2 less than 3, 2 is 2 less than 4, 3 is 2 less than 5 and 4 is 2
less than 6. The fact that these differences are all the same should help you
on your way to an answer.

Tip for Problem 16.22:
If a number is not divisible by 5, the remainder it leaves on division by 5
must be at least 1.

Tip for Problem 16.23:
The primes between 10 and 20 are 11, 13, 17 and 19. You can try out all
the possible options, but a lot of the combinations can be eliminated by
thinking about which number can be even and odd in the given context.

Tip for Problem 16.24:
If the product of three numbers is not divisible by any prime greater than
three, this must also be true of the three numbers. This greatly limits your
choice of primes or perfect squares.

Tip for Problem 16.25:
An in-law of 24 must by definition be divisible by $2 + 4 = 6$. Just take a
look at all the candidates.

Tip for Problem 16.26:
It will prove to simplify things a lot if you choose a variable to denote the
value of the middle number.

Tip for Problem 16.27:
Any number that forms a club with the two given numbers must be a divisor
of their sum. This fact severely restricts the number of candidates.

Tip for Problem 16.28:
Take a look at the prime factors of one million. What does it say about the
prime factors of a number, if its last digit is a zero?

Tip for Problem 16.29:
Note that the given conditions imply that any such number must be the
product of two fairly large primes.

Tip for Problem 16.30:
Figure out where the 5 goes first, then the rest will follow.

Tip for Problem 16.31:
The largest three-digit number is 999. This number is not divisible by 99, but checking integers with hundreds-digit 9 will lead you to the largest number with this property quite quickly. You can speed things up by considering what happens if you subtract the number formed by the last two digits from the three-digit number.

Tip for Problem 16.32:
How many numbers are divisible by 5? How many by 7? How many by both?

Tip for Problem 16.33:
Even though this problem is formulated as a divisibility problem, it really has to do with the size of the numbers.

Tip for Problem 16.34:
Take a look at the product of the ones-digits of the remaining numbers in the product after removing the even numbers and the multiples of 5. A pattern should emerge fairly quickly.

Tip for Problem 16.35:
Combining the information about the remainder and the decimals will give you some inequalities that will help you find the value of b.

Tip for Problem 16.36:
You can count how many numbers there are that are divisible by 2, 3 and 11, respectively, but watch out! If you just subtract, you are counting some numbers more than once.

Tip for Problem 16.37:
Note that the numbers 5, 4, 3, 2, 1 are alternately odd and even. This is really all you need!

SOLUTIONS

Solution for Problem 16.1:
If a number is divisible by 4 and 5, it is also divisible by 20. This means that its last digits must be 0, and its tens-digit must be even. Since the number

is also supposed to be divisible by 3, the sum of its digits must also be divisible by three. Since $2 + 0 + 0 + 7 = 9$ is divisible by three, and the ones-digit of the number must be zero, the tens-digit must therefore be an even multiple of three (greater than 0, since the two-digit number at the end is to be greater than 10), and the only such digit is 6. The two-digit number at the end must therefore be 60, and the resulting number divisible by 3, 4 and 5 is 200760.

Solution for Problem 16.2:
Since the numbers we are considering are divisible by 4, they are all certainly even. The only even two-digit numbers with digits that sum to 9 are 90, 72, 54, 36 and 18. Of these, the only ones divisible be 4 are 72 and 36. We see that there are two numbers with the required properties.

Solution for Problem 16.3:
If a number is divisible by 5, its ones-digit must be either 0 or 5. Since the numbers we are considering do not have fives among their digits, their ones-digits must be 0. Since the tens-digit can be any of the ten possible digits other than 5, and the hundreds-digit can be any of the nine digits from 1 to 9 other than 5, the total number of numbers with the required property is equal to $9 \times 8 = 72$.

Solution for Problem 16.4:
If the sum of the digits of abc with $b = c$ is divisible by 7, we know that $a + 2b$ is divisible by 7. The number abc can be written as $100a + 11b$. Since

$$100a + 11b = 98a + 7b + 2a + 4b = 7 \times (14a + b) + 2 \times (a + 2b),$$

this number is then certainly also divisible by 7.

Solution for Problem 16.5:
Since the number $5y9$ is divisible by 9, we know that $5 + y + 9 = 14 + y$ must also be divisible by 9. Since $14 + 4 = 18$, the only digit y with this property is 4. Since the resulting number must therefore be 549, the original number was $549 - 266 = 283$, giving us $x = 8$. The value of $x + y$ is therefore $8 + 4 = 12$.

Solution for Problem 16.6:
If a four-digit number is to be as small as possible, it is optimal to have 1 as the thousands-digit and 0 as the hundreds-digit. Also, the tens-digit should

be as small as possible, and so we will be best off by taking the ones-digit as large as possible. The largest possible candidate for an odd ones-digit is 7, as 9 in the ones-digit would give a digit sum greater than 9. This gives us 1017 as the smallest possible number with the required property.

Solution for Problem 16.7:

Since $N = ababab$ is divisible by 6, it is both even and divisible by 3. This means that the last digit a is even, and the sum of the digits $a + b + a + b + a + b + a = 4a + 3b$ is divisible by 3. Since

$$4a + 3b = a + 3 \times (a + b),$$

this means that a is also divisible by 3. As the first digit of N, a cannot be 0, and must therefore be 6.

Since the five-digit number $babab$ is divisible by 9, but not by 18, it must be odd, and the digit b must therefore also be odd. Since it is divisible by 9, the sum of its digits is also divisible by 9, and this is the number $b + a + b + a + b = 3b + 2a$. Since we know $a = 6$, this is equal to $3b + 12$, and the only odd value of b for which this number is divisible by 9 is 5. The number N is therefore 6565656.

Solution for Problem 16.8:

The smallest five-digit number with rising digits is 12345. This number is not divisible by 12, but if we can find one starting with the same four digits that is, it will be smallest. Indeed, the number 12348 is divisible by 12, as its digits sum to 18, which is divisible by 3, and its last two digits, namely 48, are a multiple of 4, with $4 \times 12 = 48$. The number we are looking for is therefore 12348.

Solution for Problem 16.9:

Since $36 = 4 \times 9$, we are looking for a number divisible both by 4 and by 9. Since the sum of the 4 digits we are using is $2 + 0 + 1 + 7 = 10$, the number we are looking for has at least one more digit. We must, however, add 8 (or more) in order to make the digit sum divisible by 9. The smallest such sum results by using $1 + 7 = 8$. Since the smallest number will result by putting the digits 1, 0 and 1 in the highest value spots, the smallest possible number of the required type starts with 101. The number formed by the last two digits must be divisible by 4. Of the three possible numbers formed by the remaining digits 2, 7 and 7, only 772 is divisible by 4. Therefore, 101772 is the number we are looking for.

Solution for Problem 16.10:

Since $36 = 4 \times 9$, we are looking for a number divisible both by 4 and by 9. The only two-digit number divisible by 4 that is composed of the digits 5 and 8 is 88, and these must therefore be the last two digits of the number we are looking for. The sum of these two digits is 16. Since the number is to be divisible by 9, the sum of all the digits must be a multiple of 9. The smallest possible value for the sum of the digits is therefore 18, but $18 - 16 = 2$ cannot be the result of summing fives and eights. Similarly, the next smallest multiple of 9, namely 27, is not possible either, since $27 - 16 = 11$ can also not be expressed as the sum of fives and eights. This leaves us with 36 as the next option for the sum of the digits. This is possible, since we can express $36 - 16 = 20$ with four fives. The smallest number of the type we are searching for is therefore composed of four fives, followed by two eights, or 555588.

Solution for Problem 16.11:

Max will certainly win if he puts the 8 in the ones-digit in his first move. No matter how the two play after that, the resulting number will certainly be divisible by 4, because the last two digits form the number $48 = 4 \times 12$. It will also certainly be divisible by 9, since the sum of all digits from 1 to 9 is 45, and this is divisible by 9. The game will therefore end with a number divisible by both 4 and 9, and therefore by $4 \times 9 = 36$, giving Max the win.

Solution for Problem 16.12:

Since $45 = 5 \times 9$, the product must be divisible by both 5 and 9. Since 474 is not divisible by 5, we must multiply it with a number divisible by 5, and this eliminates the option 181 from contention. Also, $474 = 3 \times 158$, and since $1 + 5 + 8 = 14$ is not divisible by 3, 474 is divisible by 3, but not by 9. If we want the product to be divisible by 9, we must multiply 474 with a number divisible by 3, and as we can easily check by adding the digits, the only one of the remaining numbers divisible by 3 is 165. We see that the product 474×165 is the only one among the options on offer with the required property.

Solution for Problem 16.13:

Since $45 = 5 \times 9$, we first note that all of Dekai's numbers are divisible by 5, since their last digits are either 0 or 5. We just have to worry about the number's divisibility by 9. We know that the sum of the digits of the number must be divisible by 9. Since the zeros do not matter as far as the sum is concerned, this sum is simply 5 times the number of fives in the number. Since 5 is a prime, the smallest multiple of 5 that is divisible by 9 is $9 \times 5 = 45$. Dekai's smallest number therefore contains nine fives, interspersed with eight zeros, i.e. 50,505,050,505,050,505.

Solution for Problem 16.14:
In order for the number to be divisible by $45 = 5 \times 9$, it must first be divisible by 5. This means that the last digit must be 0 or 5. Furthermore, in order for the number to be divisible by 9, the sum of its digits must be divisible by 9. If the last digit is 0, the sum of the four digits we know is $7 + 4 + 2 + 0 = 13$, and the missing digit must then be 5, as $13 + 5 = 18 = 2 \times 9$. This yields the solution 75420. If the last digit is 5, the sum of the digits we know is $7 + 4 + 2 + 5 = 18$, and the missing digit can be either 0 or 9 as $18 + 0 = 18 = 2 \times 9$ and $18 + 9 = 27 = 3 \times 9$ are both appropriate numbers. This yields two more solutions, namely 70425 and 79425.

Solution for Problem 16.15:
Since we want the number to be divisible by $72 = 8 \times 9$, it must be divisible by both 8 and 9. A number is divisible by 8 if and only if the three-digit number at its end is. Since we are given the hundreds-digit 7 and the tens-digit 8, the only option we have for the ones-digit is 4, since $784 = 8 \times 98$ is the only three-digit number divisible by 8 that starts with the digits 7 and 8. In order for the number to be divisible by 9, we consider the sum of its digits, as usual. The sum of the digits we already know is equal to $3 + 7 + 8 + 4 = 22$. This means that 5 is the only option for the lead digit, since $22 + 5 = 27 = 3 \times 9$ holds. The only possible number with the required property is therefore 53784.

Solution for Problem 16.16:
Any two-digit number smaller than 50 has the property, that twice the number still has two digits. (Note that $2 \times 49 = 98$, whereas $2 \times 50 = 100$.) The value of y can therefore be any number from 50 to 99. Twice these numbers are the even numbers from $2 \times 50 = 100$ to $2 \times 99 = 198$. All of these numbers have the lead digit 1, and the first digit of x is therefore certainly 1.

In order to determine numbers of this type that are divisible by 81, we can note that we can write any number of the described type as

$$2 \times 100 \times y + y = 201 \times y.$$

Since $201 = 3 \times 67$ and 67 is not divisible by 3 (as the sum of its digits is $6 + 7 = 13$, and not divisible by 3), the number $201 \times y = 3 \times 67 \times y$ is divisible by $81 = 3^4$ if and only if y is divisible by $3^3 = 27$. Since we know that y must also be a number from 50 to 99, the only possible values for y are 54 and 81, yielding the numbers 10854 and 16281.

Solution for Problem 16.17:
Since 2004 is just a little bit bigger than 2000, it is useful to note that $2000 \times 15 = 30000$, and $2000 \times 20 = 40000$. We can multiply 2004 with numbers from 15 to 19 in order to get five-digit numbers starting with the digit 3. Noting that n ends in 0 also means that the number we multiply with must be divisible by 5. This means that the only option for us, is to multiply 2004 by 15, and this gets us

$$n = 2004 \times 15 = 30060.$$

The sum of the digits of n is therefore $3 + 0 + 0 + 6 + 0 = 9$.

Solution for Problem 16.18:
If the first digit of the number is 1, the leading two-digit number must be 19, since this is the only multiple of 19 that begins with the digit 1, and there is no multiple of 21 that begins with 1. In fact, no matter which digit we are given, there is always a unique way to continue the number under the given conditions, as there is always only one multiple of either 19 or 21 that begins with any digit. We note

$$1 \times 19 = 19, \quad 5 \times 19 = 95, \quad 3 \times 19 = 57, \quad 4 \times 19 = 76,$$
$$3 \times 21 = 63, \quad 2 \times 19 = 38, \quad 4 \times 21 = 84, \quad 2 \times 21 = 42,$$
$$1 \times 21 = 21, \quad 1 \times 19 = 19$$

which gives us the only possible sequence of digits with this property as

$$1, 9, 5, 7, 6, 3, 8, 4, 2, 1, \ldots .$$

There is a cycle of length 9 in the digits, since the repetition of the digit 1 forces us to continue with 9, 5, 7, 6, 3, 8, 4, 2, 1, ... and so on until the end.

Since 1998 is divisible by 9, the ones-digit of the number must be the last number in the cycle, and the ones-digit is therefore a 2.

Solution for Problem 16.19:

As in the last problem, each digit of the number forces ones specific value for the next digit, because of the uniqueness of the multiples of 13 starting with the required digits. We note

$$2 \times 13 = 26, \quad 5 \times 13 = 65, \quad 4 \times 13 = 52,$$

which gives us the only possible sequence of digits as

$$2, 6, 5, 2, \ldots.$$

There is a cycle of length three in the digits and the sum of any triple of three successive digits is $2 + 6 + 5 = 13$. The 2002-digit number is made up of $2001 \div 3 = 667$ such triples followed by a final digit 2, and the sum of all digits is therefore equal to

$$667 \times 13 + 2 = 8673.$$

Solution for Problem 16.20:

We could just do the calculations here, but it is faster to note that we can write the expression in a different way, namely

$$2012^2 - 2011 \times 2013 = 2012^2 - (2012 - 1) \times (2012 + 1)$$
$$= 2012^2 - 2012^2 + 1 = 1.$$

We see that the expression is equal to 1, and the remainder we are looking for is therefore also equal to 1.

Solution for Problem 16.21:

Since the remainders of the number n we are searching for are given as 1 after division by 3, 2 after division by 4, 3 after division by 5 and 4 after division by 6, it is useful to note that each of these remainders is two less than the respective divisor. The property is therefore equivalent to $n + 2$ being divisible by all of the divisors 3, 4, 5 and 6. Any common multiple of these numbers will therefore do the trick. The least common multiple of these divisors is 60, so 58 is the smallest possible positive integer with the required properties, but adding any integer multiple of 60 will give us another such number.

Solution for Problem 16.22:
If none of the four numbers is divisible by 5, they each leave a remainder of at least 1 after division by 5. The sum of four number greater than or equal to 1 is certainly at least 4, and cannot therefore be equal to 3, which contradicts the assumption that none of the four numbers is divisible by 5. At least one of them must therefore be divisible by 5.

The smallest example of three different positive integers, each of which leaves the remainder 1 after division by 5, is 1, 6 and 11.

Solution for Problem 16.23:
The primes between 10 and 20 are 11, 13, 17 and 19. All of these are odd, and dividing them by 2 always yields the remainder 1, which is not a prime. We must therefore divide by an odd prime. We can find all the solutions by attacking this systematically. Dividing each of the numbers by 3, 5, 7 and so on and checking the quotients and the remainders gives us the following divisions:

$$11 \div 3 = 3; \ R = 2, \quad 13 \div 5 = 2; \ R = 3, \quad 17 \div 3 = 5; \ R = 2,$$
$$17 \div 5 = 3; \ R = 2, \quad 17 \div 7 = 2; \ R = 3, \quad 19 \div 7 = 2; \ R = 5.$$

Solution for Problem 16.24:
Since the product of the three numbers is not divisible by any prime greater than three, this must also be true of each of the three numbers. The prime number can therefore only be 2 or 3. The perfect square smaller than 20 can be 1, 4, 9 or 16. Since the sum of the three numbers is greater than 30, and we know that the prime is at most 3 and the third number is at most 18 (the largest number smaller than 20 with no prime factors other than 2 and 3), the perfect square must be 16 in order for the sum to be greater than 30. This leaves only the options $2 + 16 + 18 = 36$, which is not prime, and $3 + 16 + 18 = 37$, which is. The numbers are therefore 3, 16 and 18.

Solution for Problem 16.25:
Any in-law of 24 must be divisible by $2 + 4 = 6$. We are therefore looking for all two-digit multiples of 6 with the property that the sum of their digits is a divisor of 24. The divisors of 24 are 1, 2, 3, 4, 6, 8, 12 and 24. It therefore remains to look through the list of two-digit multiples of 6 and check for cases in which the sum of the digits is included in this list. This is the case for 12, 24, 30, 42, 48, 60, 66 and 84.

Solution for Problem 16.26:

If the middle odd number is x, the smaller number is $x - 2$ and the larger one is $x + 2$. Their sum is then

$$(x - 2) + x + (x + 2) = 3x.$$

In order for this number to be both divisible by 9 and smaller than 30, x must be odd, divisible by 3, and less than 10. This gives us two options for x, namely 3 and 9. The three triples are therefore $(1, 3, 5)$ and $(7, 9, 11)$, with $1 + 3 + 5 = 9$ and $7 + 9 + 11 = 27$.

Solution for Problem 16.27:

Since the third number that forms a club with 8 and 24 must divide $8 + 24 = 32 = 2^5$, the only candidates are the powers of 2 from 1 to 32, i.e. 1, 2, 4, 8, 16 and 32. Since 24 must divide the sum of this number with 8, all of these but the last two are too small, since the sums would be less than 24. The number 16 works, since 8 divides $16 + 24 = 40$ and 24 divides $8 + 16 = 24$. Unfortunately, 32 does not work, since 24 does not divide $8 + 32 = 40$. The only number that forms a club with 8 and 24 is therefore 16.

Solution for Problem 16.28:

The number one million can be written in several ways. We can write

$$1000000 = 10^6 = 2^6 \times 5^6.$$

If both prime factors 2 and 5 are present in any number, the last digit will be 0. This means that one of the two numbers must be composed only of fives and the other only of twos. The numbers we are looking for are therefore $2^6 = 64$ and $5^6 = 15625$.

Solution for Problem 16.29:

If a number has three factors larger than 30, it must be larger than $30 \times 30 \times 30 = 27000$. Any number fulfilling the conditions of the problem must therefore be a product of two primes greater than 30. Since $45^2 = 2025$, the smaller of the two prime factors must certainly be smaller than 45. This means that the only candidates as smaller factor are 31, 37, 41 and 43.

We can now divide the upper and lower bound in each case to find possible partners for each of these. With $1900 \div 31 \approx 61.3$ and $1999 \div 31 \approx 64.5$, there are no possible partners for 31, as neither 62, 63 nor 64 are primes. With $1900 \div 37 \approx 51.4$ and $1999 \div 37 \approx 54.0$, we obtain a possible partner

53, and indeed the number $37 \times 53 = 1961$ has all the required properties. With $1900 \div 41 \approx 46.3$ and $1999 \div 41 \approx 48.8$, we obtain a possible partner 47, and indeed the number $41 \times 47 = 1927$ also has all the required properties. Finally, with $1900 \div 43 \approx 44.2$ and $1999 \div 43 \approx 46.5$, there are no further possible partners for 43, as neither 45 nor 46 are primes. We have therefore found two numbers with the required properties, namely 1927 and 1961.

Solution for Problem 16.30:
Since the middle number bcd is divisible by 5, its last digit must be the 5, and we have $d = 5$. Since the number abc is divisible by 4, the digit c must be even, and therefore either 2 or 4. Since the number cde is divisible by 3, $c + d + e$ must be divisible by 3. If we assume $c = 2$, this means that $2 + 5 + e$ must be divisible by 3, and this is not the case for $e = 1$, $e = 3$ or $e = 4$. This means that $c = 4$ must hold, and since the number bc must be divisible by 4, this implies $b = 2$, as 24 is divisible by 4, whereas neither 14 nor 34 is. Since $4 + 5 + e$ must be divisible by 3, this now implies $e = 3$, and therefore we are left with $a = 1$. The five-digit number we are looking for is therefore 12453.

Solution for Problem 16.31:
The largest three-digit number is 999. This number is not divisible by 99, but we can find a number with hundreds-digit 9 that does have the required property. If the three-digit number $9xy$ is divisible by xy, so is the number $9xy - xy = 900$. This means that xy must be a divisor of 900. The largest two-digit divisor of 900 not including the digit 0 is 75. The largest three-digit number with the required property is therefore 975, with $975 \div 75 = 13$.

Solution for Problem 16.32:
Considering all integers from 1 to 999, every fifth one is divisible by 5. This means that $\lfloor 999 \div 5 \rfloor = 199$ of the numbers have this property. Every seventh is divisible by 7, and there are therefore $\lfloor 999 \div 7 \rfloor = 142$ such numbers. Noting that there are $\lfloor 999 \div 35 \rfloor = 28$ numbers among these that are divisible by both, since we have $5 \times 7 = 35$, there are $199 + 142 - 28 = 313$ numbers among the 999 that are divisible by either 5 or 7 or both. That leaves $999 - 313 = 686$ numbers that are divisible by neither 5 nor 7 among the positive integers smaller than 1000.

Solution for Problem 16.33:

Since there are 49 positive integers involved, their average is $999 \div 49 \approx 20.4$. At least one of the numbers must therefore be smaller than 21. We now take a look at the sum of the numbers. The greatest common divisor of a group of numbers must also be a divisor of their sum. The divisors of 999 less than 21 are 1, 3 and 9 (with the next largest being 27). The smallest number can therefore not be divisibly by any other divisor of the sum 999, and 9 is therefore the largest possible value for the greatest common divisor. This value is, indeed, possible. For instance, we can choose the number 9 as the value of 48 of the numbers and 567 as the value of the 49th.

Solution for Problem 16.34:

Once we have removed the even numbers and the numbers divisible by 5, the product of the ones-digits of the remaining numbers is

$$9 \times (1 \times 3 \times 7 \times 9) \times (1 \times 3 \times 7 \times 9) \times \cdots \times (1 \times 3 \times 7 \times 9),$$

with the expression in brackets repeated 8 times. The ones-digit of each such expression $1 \times 3 \times 7 \times 9$ can easily be calculated as being equal to 9. The ones-digit of the product is therefore equal to the ones-digit of the ninth power of 9, and since the ones-digit of $9 \times 9 = 81$ is 1, this is equal to 9.

Solution for Problem 16.35:

Since the division yields 87 with a remainder of 10, we have $a = 87b + 10$. Since division of the remainder 10 by b yields $0.43 \ldots$, we obtain

$$\frac{43}{100} \leq \frac{10}{b} < \frac{44}{100},$$

and this can be rewritten as

$$b \leq \frac{1000}{43} < 24 \quad \text{and} \quad b > \frac{1000}{44} > 22.$$

The only possible value for b is therefore 23, and this also gives us $a = 87 \times 23 + 10 = 2011$.

Solution for Problem 16.36:

In the first step, we eliminate all 50 even numbers, leaving 50. In the next step, we eliminate all off multiples of 3. There are $\lfloor 100 \div 3 \rfloor = 33$ multiples of three among the numbers from 1 to 100, and these are alternately odd and even, starting with the odd number 3. There are therefore 17 odd numbers of this type, and erasing those leaves $50 - 17 = 33$ numbers on the blackboard. In the last step, we erase any multiples of 11 that are still left. These are 11, 55 and 77, and this means that there are $33 - 3 = 30$ numbers left on the blackboard at the end.

Solution for Problem 16.37:

First of all, we will consider the reason for the impossibility of cutting in the order 5, 4, 3, 2, 1. Ignoring the first step, we note that cutting 4 (in the second step) is only possible if we cut from an even number, with the result then also being even. If we then cut the odd number 3 (in the third step), the result will certainly be odd. This means that we will certainly not be able to cut 2 in the next (fourth) step, as the number will not be divisible by 2.

Finally, in order to find the smallest number from which we can cut 1, 2, 3, 4, 5 in some order, we note that the result of cutting can never be negative. This means that the smallest number we can hope to cut these five numbers from must be $1 + 2 + 3 + 4 + 5 = 15$. It is possible to do this from 15:

$$15 - 3 = 12; \quad 12 - 4 = 8; \quad 8 - 2 = 6; \quad 6 - 1 = 5; \quad 5 - 5 = 0.$$

Chapter 17

Cutting and Tiling

Here are some fun problems concerning ways that shapes can — or can't — be cut up in regular ways. You may even want to try cutting out some of these shapes from paper and playing with them that way.

PROBLEMS

Problem 17.1:
The floor of a rectangular room is covered in 14 rows of 9 identical square tiles. Some of these tiles border the wall that surrounds the room. What fraction of the tiles does not border a wall?

(StU-05-A4)

Problem 17.2:
A tiler is laying out a floor with tiles of three types, as shown.

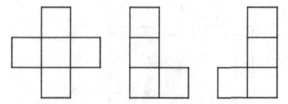

Some are composed of five unit squares in the shape of a cross, and some are composed of four in the shape of an L. Show how rectangular floors measuring 8×3 and 6×5 can be tiled. Prove that any floor measuring $6 \times k$ (with $k \geq 9$) can be covered with such tiles without cutting or letting them overlap.

(StU-95-10)

Problem 17.3:
We are given a 7 × 7-square chessboard and a number of tiles measuring 2 × 3 and 1 × 4. We want to place tiles on the board in such a way that the tiles do not overlap and do not extend beyond the edges of the chessboard. Explain why it is not possible to cover the entire chessboard under these conditions. Determine a placement in which the largest possible number of squares of the chessboard is covered. How many squares will certainly remain uncovered, if we are only allowed to use the same number of 2 × 3 and 1 × 4 tiles? Determine such a placement.

(StU-00-B3)

Problem 17.4:
Show how a 4 × 4 square can be cut into 3 × 1 rectangles in such a way that one single 1 × 1 square is left over. Show that this is also possible for a 5 × 5 square. Can a 19 × 19 square be cut into 3 × 1 rectangles and a single 1 × 1 square in a similar way? If so, describe how; if not, give reasons why not.

(StU-06-B1)

Problem 17.5:
We are given a piece of cardboard that is divided into 20 unit squares as shown. We wish to cut it along grid lines in such a way that it is split up into four congruent parts. How many of the eight light gray cells adjacent to the dark gray cell will belong to the same part as the black cell?

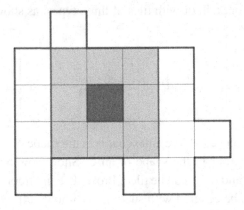

(StU-11-A4)

Problem 17.6:

A rectangle with perimeter 110 cm can be divided into five congruent rectangles, as shown. What is the area of the big rectangle?

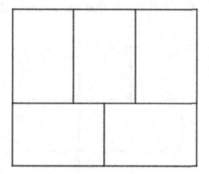

(StU-97-A2)

Problem 17.7:

The L-shaped area in the figure is to be cut into two pieces by a line through P in such a way that the area of one part is twice the area of the other part. Which of the points A, B, C, D or E can lie on the line?

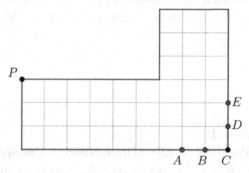

(S5K-11-10)

Problem 17.8:
A cross-shaped figure, made up of five squares whose sides are each 3 cm long, is to be cut into three pieces, each of which has the same area. Determine one way to do this and prove that the areas of the three pieces are equal.

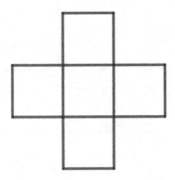

(StU-96-B5)

Problem 17.9:
Is it possible to divide a square into five isosceles triangles without leaving anything over? If so, draw such a division. If not, explain why it is not possible.

(O-01-2)

TIPS

Tip for Problem 17.1:
The number of tiles along the edge of the rectangle is almost equal to the sum of the tiles on each edge, but watch out! You may be counting something twice!

Tip for Problem 17.2:
Try creating building blocks. You can tile a 6×4 area, a 6×5 area and a 6×6 area.

Tip for Problem 17.3:
The squares on a chessboard are usually colored, with black and white squares alternating. What do you know about the number of white squares

and the number of black squares on the board and the number that can be covered by each tile?

Tip for Problem 17.4:
You might want to try your hand at the 5×5 case first. Don't be afraid of some experimentation!

Tip for Problem 17.5:
If we divide something composed of 20 cells of equal size into four parts of equal size, each part must be made up of five cells.

Tip for Problem 17.6:
Note that the top edge of the rectangle is made up of three of the shorter sides of the small rectangles, but the bottom edge is made up of two of the longer sides.

Tip for Problem 17.7:
Count the number of squares the shape is made of. If one piece is twice as large as the other, that means the smaller piece has an area one-third that of the whole shape.

Tip for Problem 17.8:
The area of each piece must be one-third of $5 \times 3^2 = 45 \, \text{cm}^2$, or $15 \, \text{cm}^2$. There are lots of ways to do this!

Tip for Problem 17.9:
Add a diagonal to the square. This already gets you halfway there!

SOLUTIONS

Solution for Problem 17.1:
Of the $14 \times 9 = 126$ tiles, those along the edge all border a wall. The total number of these is equal to the number along each edge, minus 4, since we would otherwise count the corner tiles twice. The total number of such edge tiles is therefore equal to $14 + 14 + 9 + 9 - 4 = 42$. The fraction of tiles not bordering a wall is therefore equal to $\frac{126-42}{126} = \frac{84}{126} = \frac{2}{3}$.

Solution for Problem 17.2:

First of all, here are possible solutions for the 8×3 case and the 6×5 case.

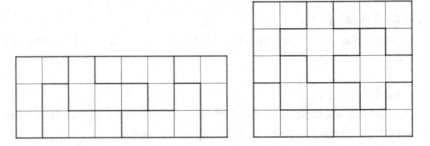

Beside the 6×5 rectangle, it is also possible to tile the 6×4 and the 6×6 rectangles, as we can see in the following figure:

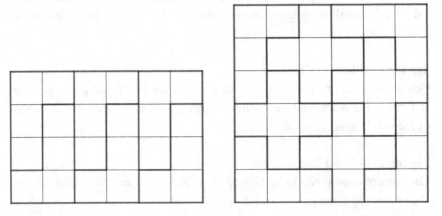

We now know that we can tile a 6×9 rectangle by joining a 6×5 and a 6×4 rectangle. Similarly, a 6×10 results by joining two 6×5 rectangles, a 6×11 results by joining a 6×5 and a 6×6 rectangle and a 6×12 results by joining two 6×6 rectangles. This means that we can tile any $6 \times k$ rectangle with $k \geq 9$, since any such a number k can be written either as

$$k = 9 + n \times 4 \quad \text{or} \quad k = 10 + n \times 4 \quad \text{or} \quad k = 11 + n \times 4 \quad \text{or}$$

$$k = 12 + n \times 4,$$

for some non-negative integer value of n.

Solution for Problem 17.3:

If we consider the squares on the chessboard to be alternately black and white, as shown, we see that it is made up of $7 \times 7 = 49$ squares, 24 of which must be in one color and 25 in the other. Any placement of a 2×3 tile will always cover three white squares and three black squares. Similarly, any placement of a 1×4 tile will always cover two white squares and two black squares. The number of black and white squares covered by any combination of tiles must therefore always cover an equal number of black and white squares. This makes a complete tiling of an unequal number of black and white squares, as is certainly the case on a 7×7 chessboard, impossible.

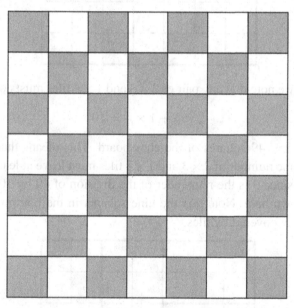

There are many ways to tile the chessboard leaving only one square uncovered. One is shown here. Note that the center square is uncovered.

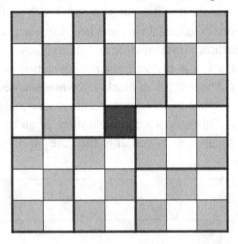

Finally, we note that any pair of 2×3 and 1×4 tiles must cover exactly

$$2 \times 3 + 1 \times 4 = 10$$

of the $7 \times 7 = 49$ squares of the chessboard. This means that any tiling using the same number of 2×3 and 1×4 tiles must leave at least 9 squares uncovered, since 9 is the remainder of the division of 49 by 10. One such tiling is shown here. Note that the nine squares in the bottom right-hand corner are not covered by tiles.

Solution for Problem 17.4:

Here are possible solutions for the 4×4 and the 5×5 square:

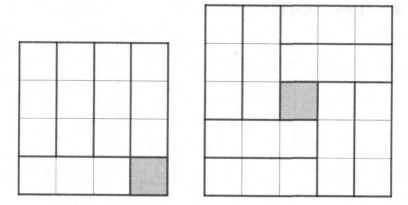

The 19×19 square is also possible. Note that you can divide it into an 18×18 square and an L-shaped part. The 18×18 square can be divided into 3×3 squares (as 18 is a multiple of 3), and each of these can certainly be cut into three 1×3 parts. The L-shaped edge is made up of two 1×18 strips, each of which can be cut into six 1×3 parts and one corner square, which will be left over.

Solution for Problem 17.5:

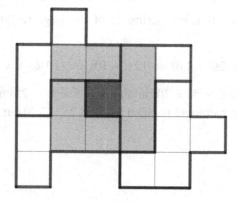

Since the cardboard figure is made up of 20 unit squares, each of the four resulting pieces must be made up of $20 \div 4 = 5$ unit squares. These pieces must therefore be congruent pentominos. This only way to cut the cardboard into five congruent pentominos is shown in the figure. We see that three of the light gray cells are on the same piece as the dark gray cell.

Solution for Problem 17.6:

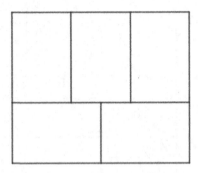

Let us say that the top and bottom edges of the large rectangle are of length a and the side edges are of length b.

Note that the top edge of the rectangle is made up of three of the shorter sides of the small rectangles, while the bottom edge is made up of two of the longer sides. We can therefore say that each short side of the small rectangles is of length $2x$, while each long side is of length $3x$. We then obtain

$$a = 6x, \quad \text{and} \quad b = 2x + 3x = 5x.$$

Since we are given that the perimeter of the large rectangle is equal to 110 cm, we have

$$2a + 2b = 110 \Leftrightarrow 12x + 10x = 110 \Leftrightarrow x = 5.$$

This gives us $a = 6 \times 5 = 30$ cm and $b = 5 \times 5 = 25$ cm, and therefore the area of the big rectangle is equal to $30 \times 25 = 750 \, \text{cm}^2$.

Solution for Problem 17.7:

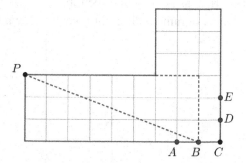

The whole shape is made up of 36 squares. Since the smaller part is supposed to be half as large as the larger one, its area must be one-third of the total area of the shape. This means that its area must be equal to $36 \div 3 = 12$ squares. Connecting P to B yields a right triangle with the area $\frac{1}{2} \times 8 \times 3 = 12$. As illustrated in the figure, this triangle is half of a rectangle made up of 24 squares, and there are 12 more squares not part of this rectangle.

Solution for Problem 17.8:

There are several ways to do this. One idea is to divide each of the squares into 3 equal strips each, or $3 \times 5 = 15$ altogether. That makes it possible to divide the whole shape into three pieces, each of which is composed of five strips. One solution of this type looks like this:

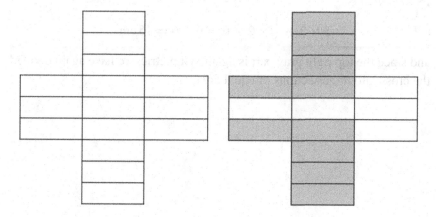

As we can see in the right figure, each of the two gray sections is composed of five such strips, as is the white section.

Two more solutions of quite a different type are shown here:

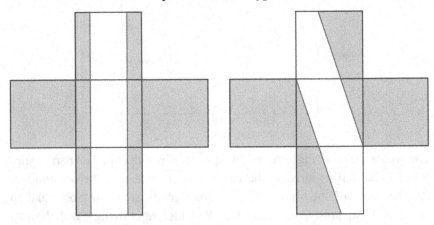

In the figure on the left, the gray section on the left is composed of a $3\,\text{cm} \times 3\,\text{cm}$ square and a $\frac{2}{3}\,\text{cm} \times 9\,\text{cm}$ rectangle. Its total area is

$$3 \times 3 + \frac{2}{3} \times 9 = 9 + 6 = 15\,\text{cm}^2,$$

or one-third of the total area of the cross shape. Since the gray area on the right has the same area, the cross is divided into three sections of equal area.

In the figure on the right, the bottom-left gray pat is composed of a $3\,\text{cm} \times 3\,\text{cm}$ square and a right triangle with sides of length $2\,\text{cm}$ and $6\,\text{cm}$. Its total area is

$$3 \times 3 + \frac{1}{2} \times 2 \times 6 = 9 + 6 = 15\,\text{cm}^2,$$

and since the top right gray part is again symmetric, we have again divided the cross into three sections of equal area.

Solution for Problem 17.9:

As we can see in the figure, this is certainly possible. The example shown has one large right isosceles triangles and four smaller congruent right isosceles triangles, but it is also possible to solve this in different ways. See how many you can find!

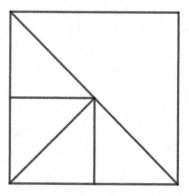

Chapter 18

Sequences

A lot of interesting things can happen when we put things in a row according to some underlying mathematical principle. When we put numbers in a row in this way, we get a sequence. Here are some enjoyable problems concerning sequences of positive integers.

PROBLEMS

Problem 18.1:
Determine the value of

$$2003 - 2002 + 2001 - 2000 + 1999 - \cdots - 4 + 3 - 2 + 1.$$

(StU-03-A2)

Problem 18.2:
We are given

$$n = 1 + 3 + 5 + 7 + \cdots + 2001 \quad \text{and}$$
$$m = 2 + 4 + 6 + 8 + \cdots + 2000.$$

Determine the value of $n - m$.

(StU-01-A8)

Problem 18.3:
The sum of eight successive odd positive integers is 112. Determine the smallest and largest of these integers.

(O-01-1)

Problem 18.4:

Which of the following numbers is larger?
$$x = 1996 \times (1 + 2 + 3 + \cdots + 1996 + 1997) \text{ or}$$
$$y = 1997 \times (1 + 2 + 3 + \cdots + 1995 + 1996).$$

(StU-97-B2)

Problem 18.5:

In the expression

$$100 \times 99 - 99 \times 98 - 98 \times 97 + 97 \times 96 + 96 \times 95 - 95 \times 94$$

$$- 94 \times 93 + \cdots + 4 \times 3 - 3 \times 2 - 2 \times 1 + 1 \times 0,$$

the order of the operation signs is $+ - - + + - - + +$, and so on.
Determine the value of the expression.

(StU-12-A10)

Problem 18.6:

Several consecutive positive integers are written on a blackboard in ascending order. The smallest number is the difference of the 13th and the 7th number, the largest number is the sum of the 13th and the 7th number. Determine the smallest and the largest number on the blackboard.

(StU-10-B5)

Problem 18.7:

The integers from 1 to 99 are written in order, forming the number

$$n = 12345678910111213 14 \ldots 9899.$$

How many digits does n have? Twenty digits are removed from n in such a way that the resulting number z is as large as possible. How often does the digit 1 occur in z?

(StU-05-B4)

Problem 18.8:

Joe and Schmo are sitting on the shore counting pebbles. Joe counts them in the usual way as $1, 2, 3, 4, 5, 6, 7, \ldots$. Schmo leaves out all the numbers that are written using the digit 3. This means that Joe's count of 1, 2, 3, 4 gives Schmo 1, 2, 4, 5, and later on Schmo will also count

..., 28, 29, 40, 41, 42, 44, 45, ... There are exactly 100 pebbles on the beach. Which number will Schmo say when Joe reaches 100?

(StU-16-A9)

Problem 18.9:

The residents of the planet Toqq each have three hands with 2 fingers apiece. Because of this, they do not count

$$1, 2, 3, 4, 5, 6, 7, 8, 9, 10, 11, \ldots, 98, 99, 100, \ldots$$

as we do, but rather

$$1, 2, 3, 4, 5, 10, 11, 12, 13, 14, 15, 20, \ldots, 54, 55, 100, \ldots$$

An Earth-man and a Toqq-woman want to get married, and while they are writing the wedding invitations, they notice that the number of wedding guests they are inviting is a two-digit number ending in 0 in both counting systems. How many guests are they inviting to their wedding?

(O-98-5)

Problem 18.10:

The sequence

$$1, 5, 4, 8, 7, 11, 10, \ldots$$

is created by the following rules. The first number is 1 and each consequent number results by alternately adding 4 and subtracting 1 from the previous one. Determine the 2003rd number in this sequence. Does the number 2003 appear as an element of the sequence? If yes, how many elements of the sequence appear before it? If no, why not?

(StU-06-B4)

Problem 18.11:

The numbers $a_1 = 1$, $a_2 = 1$ and $a_3 = 2$ are the first, second and third number in a sequence. In this *Fibonacci sequence*, every number, beginning with the third one, is the sum of the two preceding numbers:

$$a_3 = a_1 + a_2 = 1 + 1 = 2, \ a_4 = a_2 + a_3 = 1 + 2 = 3,$$

$$a_5 = a_3 + a_4 = 2 + 3 = 5, \ldots$$

Is the 2006th Fibonacci-number a_{2006} odd or even?

(StU-03-B4)

TIPS

Tip for Problem 18.1:
You could do the calculation by hand, but it would take a while. If you look at the alternating nature of the numbers being subtracted and added, you should notice something quite helpful.

Tip for Problem 18.2:
Take another look at the previous problem. A little bit of rearrangement will let you use the same trick here.

Tip for Problem 18.3:
What must the sum of the smallest and the largest of the numbers be?

Tip for Problem 18.4:
The sum s of n successive numbers can be determined in the following way. Taking the sum twice, note that the first and last number, the second and the next-to-last number, and so on, always add up to the same value k. Therefore, we obtain $2s = n \times k$.

Tip for Problem 18.5:
Consider the expression in groups of four.

Tip for Problem 18.6:
Introduce a variable for the value of the smallest number and express the 7th and 13th numbers in terms of that variable.

Tip for Problem 18.7:
How many one-digit numbers are there in the sequence used to create n? How many two-digit numbers? Which digits will we need to be in the front slots for the number z to be as large as possible after we have removed the appropriate 20 digits from n?

Tip for Problem 18.8:
Note that Schmo leaves out all the numbers with ones-digit 3 and all the numbers with tens-digit 3. How many of those are there? How many more does he need to reach Joe's 100?

Tip for Problem 18.9:
What does it mean if a number in the Toqq-system ends in a zero?

Tip for Problem 18.10:
This problem becomes much easier if you consider two separate sequences, namely the one formed by the numbers in the odd slots (the first, the third, the fifth, ...) and the one formed by the numbers in the even slots (the second, the fourth, the sixth, ...).

Tip for Problem 18.11:
Take a look at the parity (even – odd) of the first few numbers in the sequence. Do you notice a pattern?

SOLUTIONS

Solution for Problem 18.1:
If we insert parentheses in the given expression in the following way:

$$(2003-2002)+(2001-2000)+(1999-1998)+\cdots+(5-4)+(3-2)+1,$$

we see that the value of the expressions in each of the parentheses is 1. The total number of ones is equal to $2004 \div 2 = 1002$, and this is therefore the value of the expression.

Solution for Problem 18.2:
Taking a close look at the numbers involved here gives us a clue that the following rearrangement of the numbers might be useful.

$$\begin{aligned} n - m &= (1+3+5+7+\cdots+2001) - (2+4+6+8+\cdots+2000) \\ &= 1+3+5+7+\cdots+2001 - 2 - 4 - 6 - 8 - \cdots - 2000 \\ &= 1+(3-2)+(5-4)+(7-6)+\cdots+(2001-2000) \\ &= 1+1+1+1+\cdots+1 \\ &= 1001. \end{aligned}$$

Solution for Problem 18.3:
This can be solved by standard algebraic methods, but it is much easier if you think about it a bit. The smallest and largest of the numbers must have the same sum as the second smallest (two more than the smallest) and the second-largest (two less than the largest). This must also be same as the sum of the next two and of the middle two. There are four such pairs, so the sum in each pair must be $112 \div 4 = 28$. Since the largest of the eight is seven

steps of two higher than the smallest, it is larger by 14. This means that the smallest number must be $(28 - 14) \div 2 = 7$ and the largest $7 + 14 = 21$. The whole sequence of numbers is

$$7, 9, 11, 13, 15, 17, 19, 21.$$

Solution for Problem 18.4:

The sum s_n of the first n successive numbers can be determined in the following way. Taking the sum twice, we note that the first and last number, the second and the next-to-last number, and so on, always add up to the same value k. In other words, we have

$$k = 1 + n = 2 + (n - 1) = 3 + (n - 2) = \cdots = n + 1.$$

This means that we can calculate twice the sum as

$$2s_n = n \times (n + 1), \quad \text{or} \quad s_n = \frac{n(n + 1)}{2}.$$

We therefore obtain

$$1 + 2 + 3 + \cdots + 1997 = \frac{1997 \times 1998}{2} \quad \text{and}$$

$$1 + 2 + 3 + \cdots + 1996 = \frac{1996 \times 1997}{2}.$$

From this, we obtain $x = \frac{1996 \times 1997 \times 1998}{2}$ and $y = \frac{1997 \times 1996 \times 1997}{2}$, and x is therefore obviously larger than y.

Solution for Problem 18.5:

We can divide the given expression into 25 groups of the form

$$(n + 4) \times (n + 3) - (n + 3) \times (n + 2) - (n + 2) \times (n + 1) + (n + 1) \times n.$$

Multiplying each of these groups out gives us

$$(n + 3) \times ((n + 4) - (n + 2)) - (n + 1) \times ((n + 2) - n)$$

$$= 2 \times (n + 3) - 2 \times (n + 1) = 4.$$

Since there are 25 such groups, the value of the given expression is $25 \times 4 = 100$.

Solution for Problem 18.6:
If the smallest of the consecutive integers is x, the 7th number is $x + 6$ and
the 13th number is $x + 12$. Since we are given that the smallest number is
equal to the difference of the 13th and the 7th number, we have

$$(x + 12) - (x + 6) = x,$$

or $x = 6$. This means that the 7th number is $6 + 6 = 12$ and the 13th number
is $6 + 12 = 18$. Since the largest number is the sum of the 13th and the 7th
number, the largest number is therefore $12 + 18 = 30$.

Solution for Problem 18.7:
Since the number

$$n = 12345678910111121314\ldots9899$$

is composed of the nine single-digit numbers from 1 to 9 and the ninety
two-digit numbers from 10 to 99, the total number of digits in n is equal
to $9 + 2 \times 90 = 189$. If we want the number z to be as large as possible,
its first digit must be a 9. We must therefore certainly eliminate the first 8
digits, namely the digits from 1 to 8. In order to make the next digit as large
as possible, we must eliminate the smallest 12 among the next 13 digits.
The next 13 digits after the first 9 are

$$1011121314151.$$

The largest of these digits is the 5, and we therefore eliminate the digits
10111213141 before the 5 and the 1 after the 5. This means that the number
z begins with 956, and then continues with the same digits as n, starting
with 17181920…and continuing to …9899. The number of ones remaining
among these is therefore $3 + 8 = 11$, as there are three ones in the tens-digits
of 17, 18 and 19 and eight ones in the ones-digits of $21, 31, \ldots, 91$.

Solution for Problem 18.8:
First of all, let us take a look at the numbers that Schmo leaves out. He drops
all 10 numbers with the tens-digit $3(30, 31, 32, 33, \ldots, 39)$ and all ten with
the ones-digit $3(3, 13, 23, 33, \ldots, 93)$. Both groups include the number 33,
and this means that Schmo leaves out a total of $10 + 10 - 1 = 19$ numbers
up to and including the number 100. This means that he still has 19 more
pebbles to count when he has reached 100. Of the numbers 101 to 119, there

are two that include the digit 3, namely 103 and 113, which he also leaves out. He therefore finishes his counting with the number 120 and 121, and 121 is the number Schmo says when Joe reaches 100.

Solution for Problem 18.9:
Note that the numbers in the Toqq-system that end in 0 are always the multiples of 6. The Toqq number 10 is our 6. Their 20 is our 12. This means that a number that will be written with a 0 at the end in both systems must be divisible by 6 and by 10. The smallest number with this property is 30, which is written in the Toqq-system as 50, because $30 = 5 \times 6$.

Solution for Problem 18.10:
Due to the definition of this sequence, it is quite useful to take a look at two sub-sequences, namely the sequence of numbers with odd indices and the sequence of numbers with even indices. The first of these is

$$a_1, a_3, a_5, a_7, \ldots = 1, 4, 7, 10, \ldots,$$

and the second is

$$a_2, a_4, a_6, a_8, \ldots = 5, 8, 11, 14, \ldots.$$

In both of these, each number in the sequence is by three greater than the preceding one. In the case of the odd indices, we obtain each number by first adding 4 and then subtracting 1, which results in a net addition of 3. In the case of the even indices, it is the other way around. Here, we obtain each number by first subtracting 1 and then adding 4, which also results in a net addition of 3. The first sequence is therefore composed of all numbers that leave a remainder of 1 after division by three, and the second is composed of all numbers (starting with 5) that leave a remainder of 2 after division by three. For odd indices n, we can write

$$a_n = 1 + 3 \times \frac{n-1}{2},$$

and for even indices n, we can write

$$a_n = 2 + 3 \times \frac{n}{2}.$$

Since 2003 is odd, we see that the 2003rd number in the sequence is

$$a_{2003} = 1 + 3 \times \frac{2003 - 1}{2} = 1 + 3 \times 1001 = 3004.$$

Furthermore, the number 2003 does, indeed occur in the sequence, as it leaves a remainder of 2 after division by three. It is therefore in the sub-sequence of numbers with even indices, and we have

$$a_n = 2003 = 2 + 3 \times \frac{n}{2},$$

which yields

$$2001 = 3 \times \frac{n}{2} \Leftrightarrow 1334 = n.$$

We see that there are 1333 elements in the sequence before the appearance of 2003.

Solution for Problem 18.11:
Note that the first two numbers $a_1 = 1$ and $a_2 = 1$ are odd. Adding two odd numbers gives us an even result, and therefore, even without doing the actual calculation, we see that $a_3 = a_1 + a_2$ must be even. Adding an even number and an odd number gives an odd result, and therefore $a_4 = a_2 + a_3$ must be odd, since a_2 is odd and a_3 is even. For the same reason, $a_5 = a_3 + a_4$ is odd, too. We see that this produces a cycle that repeats indefinitely:

$$\text{odd, odd, even, odd, odd, even, } \dots,$$

with every third number in the sequence being even. In other words, since the even numbers start with a_3, the numbers in the sequence with indices divisible by 3 are even, and all the others are odd. Since 2006 is not divisible by three, the Fibonacci-number a_{2006} must be odd.

Chapter 19

Counting Things

A large number of interesting questions arise just by asking how many things can have a certain property or how many ways there are to do something. In this section, we will take a look at a number of problems of this type. Some are quite abstract, dealing with number digits or points and lines in the plane. Others are quite practical. In how many ways can we sit on a bench, if we consider a number of restrictions? How many rooms are there in a certain building if we know something specific about the rooms?

PROBLEMS

Problem 19.1:
The page numbers of a book are written using a total of 642 digits. How many pages does the book have?

(O-93-1)

Problem 19.2:
After my dog took today's newspaper on a trip all around the house, the pages were all over the place. While I was picking up the sheets, I noticed that pages 6 and 20 were printed on the same sheet. What were the other two pages printed on this sheet?

(O-94-6)

Problem 19.3:
Verena is mad at her brother Martin and decides to tear up his favorite book. First, she tears the cover off. Now she can see page 1 in the front and page 120 in the back. Then, she tears the first 10 sheets and the last two sheets

off. As Martin returns to his room, the remaining piece is open to its middle. Which page is visible on the left side?

(S5K-07-10)

Problem 19.4:
There are 7 blue balls, 9 red balls and 10 white balls in a container. Balls are taken out of the container without looking at them, until either three balls of the same color have been taken out, or one ball of each color. What is the largest number of balls that must be taken out for this to be the case?

(O-94-7)

Problem 19.5:
A box contains less than 1000 balls. Exactly one-third of the balls are red, one-quarter are blue and the rest are green. What is the largest number of balls that could be in the box under these circumstances?

(S5K-02-1)

Problem 19.6:
Astrid goes to the county fair and buys a packet of chocolate sticks. The packet contains four different types of sticks: five each are milk chocolate and coconut, four are rum chocolate and four are rum coconut. She wants to eat a stick containing coconut first, but she can't tell what kind she has in her hand until she takes it out of the packet. How many does she have to take out of the packet so that she is sure to have one with coconut among them?

(O-02-5)

Problem 19.7:
Ms Schlamperl has a box in her attic containing various woolen mittens that cannot be differentiated by touch. Three are left red mittens, six are left green mittens, four are right red mittens and six are right green mittens. What is the smallest number of mittens she has to take out of the box in order to be sure of having a left and a right mitten of the same color?

(StU-98-A4)

Problem 19.8:
A deck of cards is made up of 28 cards. There are seven values (8, 9, 10, jack, queen, king and ace) in each of the four suits diamonds, spades, hearts

and clubs. The deck is shuffled thoroughly. At least how many cards must I draw in order to be certain of holding either a heart card or a king?

(StU-17-A7)

Problem 19.9:
A quadrilateral has two diagonals. A pentagon has 5 diagonals. How many diagonals does an octagon (a figure with 8 corners) have? (A diagonal is the line segment joining two non-adjacent corners.)

(StU-17-A10)

Problem 19.10:
If we draw three lines in the plane, we obtain at most three points of intersection. What is the largest number of points of intersection we can obtain if we draw six different lines in the plane?

(S5K-06-1)

Problem 19.11:
Is it possible to draw 10 lines in the plane in such a way that they determine exactly 19 points of intersection? Is it possible with exactly 98 points of intersection?

(StU-98-B1)

Problem 19.12:
If four lines in the plane are parallel (as shown in the figure on the left), no two of these lines have a common point. If the four lines lie as generally as possible (as shown in the figure on the right), there are six points that lie on two of the lines each. Is it possible for us to draw six lines, such that exactly four points lie on (at least) two of the lines? Is it possible for there to be exactly five?

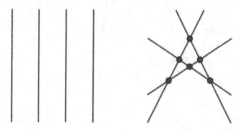

(StU-01-B1)

Problem 19.13:
Each corner of a square $ABCD$ is colored either red or green. The square must have the same number of red and green corners. In how many different ways can the square be colored under these rules?

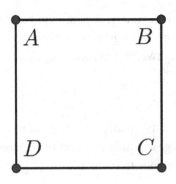

(StU-16-A6)

Problem 19.14:
A square $ABCD$ is drawn on a piece of paper. How many ways are there to draw a second square on the same piece of paper, such that the new square has exactly two corners in common with $ABCD$?

(S5K-07-4)

Problem 19.15:

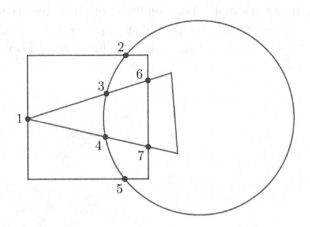

In the figure we see a circle, a square and a triangle placed in such a way that a total of 7 points of intersection result. What is the largest number of points of intersection that can result by placing a circle, a square and a triangle in an appropriate manner?

(S5K-11-5)

Problem 19.16:
Isosceles triangles are triangles in which at least two sides have the same length. (This means that triangle with three sides of equal length are also isosceles.) How many different isosceles triangles are there, if the sides of the triangles all have lengths of either 2 cm, 3 cm, 7 cm or 11 cm. (Note that triangles with three pairs of sides with pairwise equal lengths do not count as different.)

(S5K-13-6)

Problem 19.17:
How many different isosceles triangles with perimeter 23 cm exist, in which the lengths of all sides (in cm) are positive integers?

(StU-06-A4)

Problem 19.18:
How often do we write the digit 6 if we write all the integers from 0 through 2006?

(S5K-06-7)

Problem 19.19:
A number like 424 or 167761, that reads the same from left to right and from right to left, is called a *palindrome*. How many even four-digit numbers are palindromes?

(StU-02-A4)

Problem 19.20:
How many four-digit numbers exist, in which the digits 1, 2, 3 and 6 all appear? How many of them are divisible by four?

(StU-01-B2)

Problem 19.21:
How many positive integers smaller than 200 are there, whose decimal representation contains exactly two different digits?

(StU-08-A9)

Problem 19.22:
In a four-story hotel, the elevator is at one end of the hall and a window at the other end in each story. The room numbers all have three digits. The first digit is always the story in which the room is situated and the other two are running numbers starting with 01 left of the elevator, counting on the same side of the hall all the way to the window, and then continuing on the other side back to the elevator. Vaclav has room number 208 and right across from his door is door number 215. How many rooms are there in the hotel if all stories are the same and doors are always opposite to each other?

(O-99-6)

Problem 19.23:
The diagram on the right shows the roads in a city grid. All of the roads but one run in either a strict north–south or a strict east–west direction. Due to road works, the point marked with an X is currently blocked. Out of all the possible routes from A to B, some are the shortest. How many such shortest routes are there?

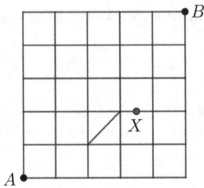

(StU-97-A6)

Problem 19.24:
To keep from being too bored while she is waiting for a bus, Andrea plays a game. She takes 2 steps forward, one step back, 2 forward, one back and

so on, until she reaches the signpost that she was 9 steps away from when she started. How many steps does she have to take in total before she reaches the signpost for the first time?

(S5K-04-3)

Problem 19.25:
Alex, Bernd and Christian are running up the stairs in their filthy sneakers. Alex takes two steps at a time, Bernd takes three steps at a time and Christian takes five steps at a time. The only step they all leave footprints on is the top one; all the others have either 0, 1 or 2 footprints. How many steps remain unsullied, with none of the kids having stepped on it?

(StU-05-A8)

Problem 19.26:
Achim, Bettina, Carl, Diana, Erich, Franz and Gerlinde need to choose a group leader for a project. In order to do this, they stand in a circle in alphabetical order and count out (with Achim following after Gerlinde, and so on). Every third person in the counting process is eliminated. The last one of the group left standing is Bettina, and she is made group leader. Which member of the group was the one who started the counting process?

(O-95-6)

Problem 19.27:
An even number of houses are placed along a circular street surrounding a pond. The house owners are planning to renovate all the houses together in order to beautify the area. This year, two of the houses are to be renovated, and in order for the work to be done in an even spread, they do not want these houses to be neighboring or immediately opposite one another across the pond. It turns out that there are 16 possible pairs of houses they could start with. How many houses are there in the street?

(S5K-14-2)

Problem 19.28:
Madame asks her gardener to plant 20 rose bushes in a circle in the middle of her garden. She wants him to use red, yellow and white roses. (In other words, all three colors must be planted somewhere.) Under no circumstances is he to plant two red rose bushes next to each other, and similarly, no white

rose bush should ever be planted next to a yellow one. We let A denote the largest number of red rose bushes that the gardener can plant under these restrictions and B the smallest such number. Determine the value of $A - B$.

(S5K-16-10)

Problem 19.29:
Six friends, Alf, Beate, Chris, Doris, Edgar and Fritz are standing in a circle to play a ball game. Beate and Doris want to stand next to each other, as do Edgar and Fritz. Unfortunately, there has been a misunderstanding in the group, and neither Edgar nor Fritz is willing to stand beside Beate or Doris. How many different arrangements are possible for the six players? (Note we say that two arrangements are the same, if every one of the friends is standing between the same two people in each of them.)

(StU-14-A6)

Problem 19.30:
Kevin, Lars, Michael, Nora and Olga want to sit in the five seats in the last row of their school bus. Nora and Olga have an iPod, and since they both want to listen to it, they want to sit together. How many different seating arrangements are possible for the five kids?

(S5K-06-5)

Problem 19.31:
Frieda, Gerda, Hannah and Ida have been walking for a while and would like to sit down on a bench that can seat exactly four people. Frieda does not want to sit on the outside under any circumstances, but Gerda definitely wants to sit on the outside. Ida has just had an argument with Gerda, and will not sit beside her. Hannah is ok with anything. How many different seating arrangements are there for the four girls on the bench?

(S5K-12-8)

Problem 19.32:
Clare is just starting to learn to play the flute. Every day she learns to play three new tones, but every night she forgets one of them again. How many days will it take until she is able to play 15 different tones in the evening?

(StU-09-A4)

Problem 19.33:
A large box is filled with six medium boxes. Some of the medium boxes are filled with six empty small boxes. Among all the boxes, there are five boxes that contain other boxes. How many of the boxes are empty?

(StU-10-A9)

Problem 19.34:
Georg is fooling around in his grandmother's attic, and discovers a box with a string tied around it. He opens the string and discovers 5 smaller boxes inside the box. Each of these contains 3 even smaller boxes, each of which contains 6 tiny boxes. At first, Georg is confused by this, but then he starts to count the boxes. How many are there altogether?

(StU-97-A1)

Problem 19.35:
In the 2a class, the pop group "Kyoto Motel" has 9 fans, and the pop group "Osaka Restaurant" has 7 fans. Till and Bom are the only two pupils in the class that like both groups, and 14 of their classmates don't like either group. How many pupils are there in the 2a class?

(S5K-06-3)

Problem 19.36:
In a group of 39 students, exactly 10 read both the books of E. Wallace and A. Christie, exactly 9 read both the books of A.C. Doyle and A. Christie, exactly 11 read both the books of A.C. Doyle and E. Wallace, exactly 6 read only the books of E. Wallace, exactly 5 read only the books of A.C. Doyle and exactly one reads only the books of A. Christie. Exactly 11 students do not read any such books. How many students read books by all three authors?

(StU-92-5)

Problem 19.37:
Three houses A, B and C are on a dead-end street. The mail carrier enters the street, delivers the mail to each house, and leaves the street again. In how many different ways can he deliver the mail in an order in which he does not have to double back? (If he delivers the mail in the order $C - A - B$, he will double back, since he has to walk twice from A to B.) How many

ways are there, if there are five houses in the street? How many if there are seven?

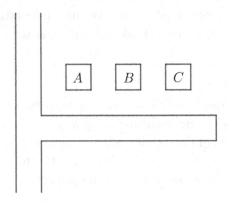

(StU-96-B6)

Problem 19.38:
13- and 14-year-olds are playing a chess tournament. Each participant plays one game against every other one. In 15 games, two 14-year-olds play each other, and in 6 games, two 13-year-olds play each other. How many games are played, in which a 13-year-old plays against a 14-year-old?

(StU-03-A9)

Problem 19.39:
An island is connected to the mainland by five bridges A, B, C, D and E. The road across bridge A is a one-way-road from the mainland to the island, the roads across bridges B and C are one-way-roads from the island to the mainland. The remaining two bridges can be used in both directions. How many different routes are there leading from the mainland to the island and back, if a route is uniquely determined by the selection of the bridge to the island and the bridge back?

(StU-08-A3)

TIPS

Tip for Problem 19.1:
How many pages use only one digit? How many use two?

Tip for Problem 19.2:
Note that pages 1 and 2 are printed on the first sheet, and pages 3 and 4 on the second. This means that pages 5 and 6 are printed on the third sheet.

Tip for Problem 19.3:
How many sheets of paper is the book printed on? How many does Verena tear off?

Tip for Problem 19.4:
It is easier to think about this backwards. What is the largest number of balls you can take out without this being the case?

Tip for Problem 19.5:
If exactly one-third of the balls are red, the total number of balls must be divisible by three.

Tip for Problem 19.6:
This is another one, in which it would be wise to think backwards. How many non-coconut sticks can she choose in the worst case?

Tip for Problem 19.7:
Once again, think backwards. What is the largest number of mittens she can take without getting a pair of the type she needs?

Tip for Problem 19.8:
A last opportunity to think backwards before we move on to other ideas.

Tip for Problem 19.9:
How many ways are there to connect each corner of the octagon to another? How many connections are there altogether? How many of these are sides of the octagon?

Tip for Problem 19.10:
If no two lines are parallel, and no three have a point in common, the number of such points can equal the number of pairs of lines.

Tip for Problem 19.11:
This problem was set in 1998. If you think about it a bit, you can see that 98 is much too large a number. On the other hand, 19 seems to be reasonable. Just remember that parallel lines don't have common points. Also, note that more than 2 lines can have a point in common!

Tip for Problem 19.12:
Again, it is helpful to note that three or more lines can pass through a common point. There are only two ways for there to be less than 6 common points. There can be parallel lines or three lines with a common point. What does this tell us if we want to have exactly 4 or 5 common points?

Tip for Problem 19.13:
There must be two red corners and two green corners. How many ways are there to choose two red corners?

Tip for Problem 19.14:
You need to watch out here. A side of one square can be a diagonal in the other!

Tip for Problem 19.15:
How many points can a line segment have in common with a circle? How many line segments make up a square?

Tip for Problem 19.16:
Note that the triangle inequality must hold, i.e. the sum of the lengths of two sides in a triangle must always be larger than the length of the third side.

Tip for Problem 19.17:
Note that the length of the longest side of a triangle cannot exceed half the perimeter.

Tip for Problem 19.18:
How often will the digit 6 be used as the ones-digit? How often as the tens-digit or the hundreds-digit?

Tip for Problem 19.19:
Choosing the last two digits already forces the first two in a four-digit palindrome.

Tip for Problem 19.20:
How many options do we have for the thousands-digit? In how many ways can we choose a hundreds-digit if we have already chosen a thousands-digit?

Tip for Problem 19.21:
The numbers in question can either be two-digit numbers containing two different digits, or three-digit numbers containing one digit twice and the

other digit once. Since these numbers are smaller than 200, their hundreds-digit must be 1.

Tip for Problem 19.22:

What are the numbers of the rooms next to the window?

Tip for Problem 19.23:

All the shortest routes must include the little diagonal. How many shortest routes are there from A to the start of this diagonal and how many are there from its end to B?

Tip for Problem 19.24:

This one is tricky! You might think that she gains one step for every two moves, for a total of 18 moves. You would be wrong, though!

Tip for Problem 19.25:

They all start at the bottom, obviously. Every step each of them hits is a multiple of the number of steps he takes at a time.

Tip for Problem 19.26:

You can do the counting process yourself. Just start anywhere in the circle, and note who is left over. Then shift your starting point to make Bettina the winner.

Tip for Problem 19.27:

Try introducing a variable here. If n is the number of houses, how many other houses can it be paired with?

Tip for Problem 19.28:

The largest possible number should be pretty obvious, but not that the smallest possible number is not 1! Why not?

Tip for Problem 19.29:

You can almost treat Beate and Doris as one person for the sake of this problem, and the same is true for Edgar and Fritz.

Tip for Problem 19.30:

This is a bit similar to the last one. Nora and Olga are almost like one person.

Tip for Problem 19.31:

There are not a lot of options for Frieda and Gerda, so you can start with them. Also, once Gerda has found her seat, there are not a lot of options left for Ida.

Tip for Problem 19.32:
You might think that she has a net gain of two a day, but it's a little bit more complicated than that. (Not much more, though.)

Tip for Problem 19.33:
How many boxes are in some other box? This information tells you how many boxes there are altogether.

Tip for Problem 19.34:
Don't be as confused as Georg! This is just a matter of keeping track of boxes of each of the four sizes.

Tip for Problem 19.35:
How many of the pupils only like the one group? How many only like the other?

Tip for Problem 19.36:
There are three authors involved here with a partly overlapping readership in the group. A diagram might help. Try drawing three overlapping circles, with each circle representing the readers of each author, and the overlap representing the readers they have in common. Then write the numbers you already know in the appropriate sections of your diagram. Or, you could try introducing variables and solving by algebra.

Tip for Problem 19.37:
It isn't hard to write down all possible options and count them.

Tip for Problem 19.38:
How many 13-year-olds are playing in the tournament? How many 14-year-olds?

Tip for Problem 19.39:
One way to solve this problem is by naming the bridges and listing all possible routes. Note that the number of possible routes back is independent of the specific bridge used to cross onto the island.

SOLUTIONS

Solution for Problem 19.1:
The first 9 pages of a book (pages 1–9) use one digit each. The next 90 (pages 10–99) use two digits each. That gives us a total of

$9 \times 1 + 90 \times 2 = 189$ digits used for numbering the first 99 pages. Since we will require 3 digits for each of the next pages, and there are $642 - 189 = 453$ digits left, there are $453 \div 3 = 151$ pages with three-digit page numbers, for a total of $99 + 151 = 250$ pages.

Solution for Problem 19.2:
Since pages 1 and 2 are on the first sheet, and pages 3 and 4 on the second, pages 5 and 6 are printed on the third sheet. Four pages are printed on every sheet between page 6 and 20, which is on the same sheet as page 6. Since there are $19 - 6 = 13$ pages between page 6 and page 20, this means that there are four sheets in between, with pages numbered 7–18. We therefore see that pages 5 and 19 are on the same sheet as pages 6 and 20.

Solution for Problem 19.3:
Since the book has pages numbered from 1 to 120, the book is printed on 60 sheets of paper. Verena tears the first 10 sides (with pages 1 through 20) and the last two (with pages 117–120) off. This leaves 48 sheets with pages 21–116. When this remaining part is flipped open in the middle, there are 24 pages on either side. The page visible on the left is the back of the 24th remaining sheet, and this is page $21 + 24 \times 2 - 1 = 68$.

Solution for Problem 19.4:
Let us take a look at the largest number of balls we can take out, without either taking one of each color of three of the same color. If we can only take two balls of the same color, we will reach the largest number by taking two of each. Four is therefore the largest number possible. If we take out five balls, and we have only taken balls of at most two colors, we must have taken at least three balls of one of these colors.

Solution for Problem 19.5:
If exactly one-third of the balls are red, the total number of balls must be divisible by three. Similarly, if one-quarter of the balls are blue, the total number of balls must be divisible by four. The total number is therefore divisible by $3 \times 4 = 12$, and the largest number smaller than 1000 and divisible by 12 is $12 \times 83 = 996$. This is possible if $996 \div 3 = 332$ balls are red, $996 \div 4 = 249$ are blue and $996 - 332 - 249 = 415$ are green.

Solution for Problem 19.6:
There are five milk chocolate sticks and four rum chocolate, for a total of $5 + 4 = 9$ non-coconut sticks in the packet. In the worst case (for her), she could take all of these out first, but she would be sure to get a coconut stick with her next choice. She therefore needs to take $9 + 1 = 10$ sticks out of the packet to be certain of getting a coconut stick.

Solution for Problem 19.7:
In the worst case, she could take all the left mittens of one color and all the right mittens of the other color without getting an appropriate pair. The largest number she can therefore take without getting a proper pair is $6 + 4 = 10$, assuming she takes all of the left green mittens and all of the right red ones. Taking an 11th mitten guarantees an appropriate pair.

Solution for Problem 19.8:
In the worst case, I could draw all of the spades, diamonds and clubs that are not kings first. There are 6 of each of these, for a total of $3 \times 6 = 18$. When I draw the 19th card, I can be certain of having at least one card that is a heart card or a king.

Solution for Problem 19.9:
Each of the eight corners of the octagon can be joined to the seven others, for a total of $8 \times 7 = 56$ connections. By counting this way, we have counted each line segment joining two corners twice, since we counted it as a connection from A to B, but also from B to A. This means that the total number of line segments joining two corners in the octagon is equal to $56 \div 2 = 28$. Of these, eight are sides of the octagon, leaving $28 - 8 = 20$ diagonals.

Solution for Problem 19.10:
Each of the six lines can have a common point with each of the five others. This would give us $6 \times 5 = 30$ possibilities, but we must note that counting this way would mean that we have identified every common point of lines a and b also as a common point of b and a. Each of these possibilities was therefore counted twice, and the total number of common points is therefore equal to $30 \div 2 = 15$.

Solution for Problem 19.11:

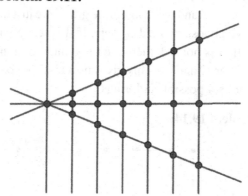

If 7 lines are parallel, and the other three intersect one of these in a common point, we have a total of $1 + 6 \times 3 = 19$ points of intersection, as shown in the figure. The number 19 is therefore possible. On the other hand, 98 is not possible. The largest number of points of intersection we can get with 10 lines is attained if any two of these have a different point in common. This means that each of the 10 lines has a common point with each of the 9 others, for a total of $10 \times 9 = 90$ pairs, or $90 \div 2 = 45$ points of intersection. The number 98 is therefore unattainable.

Solution for Problem 19.12:

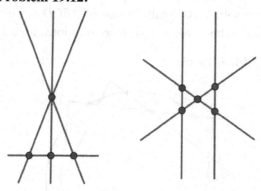

If three of the lines pass through a common point and the fourth intersects them all, there are exactly four common points in the configuration, as we see in the figure on the left. If two are parallel, and the other two intersect them both as well as each other, there are exactly five, as we see in the figure on the right.

Solution for Problem 19.13:

Since two corners are red and two corners are green, we just need to count the number of way two choose two red corners. If A is red, it can be combined with B, C or D. If A is not red, there are three more options for red pairs: B and C, B and D or C and D. This gives us a total of six options for red pairs, and therefore six possible colorings.

Solution for Problem 19.14:

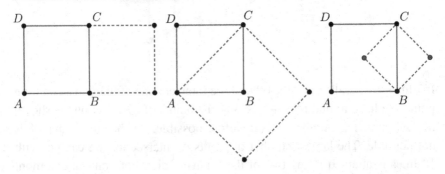

The second square can have a common side with $ABCD$, but be oriented in the opposite direction. There are four solutions of this type. Also, one of the diagonals of $ABCD$ can be a side of the new square. Since there are two diagonals, and the new square can point in two different directions, this gives us for more solutions. Finally, a side of $ABCD$ can be a diagonal of the new square, and this gives us four more solutions, for a total of 12.

Solution for Problem 19.15:

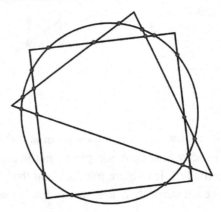

The most points a line segment can have in common with a circle is 2. Since a square has four line segments as its sides, the most points a square can have in common with a circle is equal to $4 \times 2 = 8$. Similarly, since a triangle has three sides, the most points a triangle and a circle can have in common is equal to $3 \times 2 = 6$. The same thing holds for a square: the most points a line segment can have in common with a square is 2, and the most points a triangle can have in common with a square is therefore equal to $3 \times 2 = 6$. The largest possible number of common points of two of the objects is therefore equal to $8 + 6 + 6 = 20$. A configuration with 20 such points is shown in the figure.

Solution for Problem 19.16:

Let us assume that sides a and b are of the same length. The side c may also be of the same length, but does not have to be.

If $a = b = 2$, c can be equal to 2 or 3, but not longer, as the length of c would then exceed the sum of the lengths of a and b.

If $a = b = 3$, c can be equal to 2 or 3, but not longer, as the length of c would then exceed the sum of the lengths of a and b.

If $a = b = 7$, c can be equal to 2, 3, 7 or 11.

If $a = b = 11$, c can be equal to 2, 3, 7 or 11.

This gives us a total of $2 + 2 + 4 + 4 = 12$ such isosceles triangles.

Solution for Problem 19.17:

The sum of the lengths of the equal sides in an isosceles triangle is always even. Since the perimeter of the isosceles triangles we are searching for is odd, the length of the base (in cm) must also be odd. Furthermore, the length of the longest side of a triangle cannot exceed half the perimeter. The length of the base is therefore at most 11 cm. There are six odd natural numbers less than or equal to 11. Each of these six numbers is a possible length of the base (in cm) of an isosceles triangle of the required type.

Solution for Problem 19.18:

Let us consider all numbers from 1 to 1999. Every tenth number has the ones-digit 6, so that gives us $2000 \div 10 = 200$ sixes. This is also true of the tens- and hundreds-digits, although they turn up in bunches. Still, there are 200 in each of the slots. The 6 is never used in the thousands-digit in the region we are considering. Since there is also a last digit 6 used in writing

2006 (not hard to tell when this problem was posed at a competition, is it?), the total number of sixes is therefore equal to $3 \times 200 + 1 = 601$.

Solution for Problem 19.19:

Choosing the last two digits a and b in a four-digit palindrome $baab$ already forces the first two digits. Since the first digit of a number can never be 0, and the number must be even, the value of the digit b can only be 2, 4, 6 or 8, since it is both the first and last digit in the number. The digit a can be any of the ten digits. The total number of even four-digit palindromes is therefore equal to $10 \times 4 = 40$.

Solution for Problem 19.20:

There are four options among the given digits for the thousands-digit in the four-digit number. Once we have chosen one of them, we have three options remaining for the hundreds-digit, and once we have chosen that as well, we still have two options remaining for the tens-digit. We then have no option but to choose the remaining digit as the ones-digit. This means that there exist a total number $4 \times 3 \times 2 = 24$ of such four-digit numbers. In order for one of these to be divisible by 24, the last two digits must form a two-digit number divisible by 4. The only such possibilities we can get by choosing from the given options are 12, 32, 16 and 36. Each of these can be combined with the other two remaining digits in two ways, since they can be chosen as the thousands- or hundreds digits. This gives us a total of $4 \times 2 = 8$ such numbers that are divisible by 4.

Solution for Problem 19.21:

We don't need to worry about the single-digit numbers, of course. Nearly all of the 90 two-digit numbers from 10 to 99 have two different digits. The only exceptions are the nine numbers 11, 22, 33, ..., 99, leaving a total of $90 - 9 = 81$ such two-digit numbers. Now, let's have a look at the three-digit numbers starting with 1. If the tens-digit is also 1, the ones-digit can be anything other than 1, so there are 9 such numbers. Similarly, if the ones-digit is also 1, the tens-digit can be anything other than 1, and there are also 9 numbers of this type. Finally, if neither the ones- nor the tens-digit is 1, these must both be the same, and there are again 9 choices available. The total number of integers with the required properties is therefore equal to $81 + 9 + 9 + 9 = 108$.

Solution for Problem 19.22:

If room 208 is across from room 215, the same is true for rooms 209 and 214, 210 and 213 and 211 and 212. These last two rooms must be next to the window, and since the room numbers in the second story go from 201 to 211 on one side of the hall, they must go from 212 to 222 on the other side. There are therefore 22 rooms in each story, and $22 \times 4 = 88$ in the hotel altogether.

Solution for Problem 19.23:

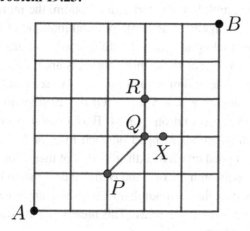

Every route from A to B will require traveling five blocks north and five blocks east in total. Any of the shortest routes will have no backward steps, and will include using the slanted road, which allows one step to make a jump both north and east. There are three ways to get from A to the point P at the south–west end of this slanted road, namely NEE, ENE and EEN. From the north-eastern point Q of the slanted road, we must then proceed north to R, as the road east is blocked. There are now four steps left to make from R to B, namely two north and two east. There are six possible orders for these steps, namely

NNEE, NENE, NEEN, ENNE, ENEN and EENN.

Since any of the three options at the bottom can be combined with any of the six at the top, we obtain a total of $3 \times 6 = 18$ shortest routes from A to B.

Solution for Problem 19.24:

At the beginning of her game, Andrea always takes 3 steps to get one step closer to the signpost, since she takes two forward steps plus one backward step. This is true until she reaches the seventh step, so she takes $7 \times 3 = 21$ steps to get there. Then, however, she only needs to take two more steps to reach the signpost, for a total of $21 + 2 = 23$.

Solution for Problem 19.25:

Since the number of each step that each boy hits is a multiple of the number he takes at a time, and they all start at the bottom, the number of the first step they will all hit again is the least common multiple of the numbers of steps each of them takes at a time. In other words, we are looking for the number $lcm(2, 3, 5)$. Since these three numbers are different primes, their least common multiple is simply their product. We see that the first step the all hit is step number $2 \times 3 \times 5 = 30$. Of the 29 steps in between, Alex has stepped on every second one, or 14. Bernd has stepped on each third one, or 9 and Christian has stepped on each fifth, or 5. Noting that Alex and Bernd both stepped on each sixth step, or 4 of them, Alex and Christian both stepped on each tenth on, or 2 and Bernd and Christian both stepped on each fifteenth, or one, the total number of steps that anyone stepped on was equal to $14 + 9 + 5 - 4 - 2 - 1 = 21$. This means that there are $29 - 21 = 8$ unsullied steps left over.

Solution for Problem 19.26:

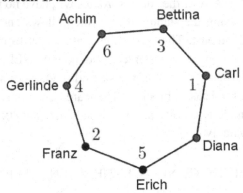

Let us assume that we start the counting process with Achim, since he is the first alphabetically. The elimination is then in the following order: Carl is eliminated first (1), then Franz (2), Bettina (3), Gerlinde (4), Erich (5) and

Achim (6), leaving Diana as the winner. (You can count this out yourself, simply take a look at the figure.) Since Bettina was actually the one who was left standing, and she is two spots to the right of Diana, counting must have been started two spots to the right of Achim, and we see that the counting process must have started with Franz.

Solution for Problem 19.27:

Let us assume that there are n houses in the street. Each house cannot be paired with a neighbor or with the house across the pond, and so there are $n - 4$ other houses it could be paired with. (Obviously, the house cannot be paired with itself, either.) The total number of pairs a house A can be paired with a house B is therefore equal to $n \times (n - 4)$, and since this method of counting counts each pair once as A with B and once as B with A, the actual number of pairs is equal to half of this, or $\frac{1}{2} \times n \times (n - 4)$. We are given that this number is equal to 16. We therefore see that there are 8 houses in the street, since

$$\frac{1}{2} \times 8 \times (8 - 4) = 16.$$

Solution for Problem 19.28:

The largest possible number is obviously attained if every other rose bush is red, and this means $A = 10$. The smallest number is attained if there is a single group of white rose bushes and a group of yellow rose bushes, and these are separated by a red one on either side. This means $B = 2$, and we obtain $A - B = 10 - 2 = 8$.

Solution for Problem 19.29:

Since the two pairs of Beate and Doris on the one hand and Edgar and Fritz on the other must be separated, Alf and Chris must stand between the two pairs. We note that either Beate or Doris can stand beside Alf from the one pair, and the same holds for either Edgar and Fritz. Choosing who from each pair stands beside Alf then determines all other positions in the circle. This means that there are only $2 \times 2 = 4$ possible arrangements of the players.

Solution for Problem 19.30:

If we treat Nora and Olga as one unit, we have four options for choosing who will sit on the very left. Having chosen that person (or that pair), we have three options to choose who will sit to their immediate right. Having made that choice, we have two choices for the next spot, and this determines the last spot. This gives us a total of $4 \times 3 \times 2 = 24$ possible arrangements of the four. Since Nora and Olga can exchange places, the total number of possible seating arrangements in the bus is therefore equal to $24 \times 2 = 48$.

Solution for Problem 19.31:

There are two options for Gerda to sit down on the outside. If Frieda sits down beside her, Ida has two options, and whichever one she chooses, Hannah will be ok with it. This gives a total of $2 \times 2 = 4$ such options. If Frieda does not sit down beside Gerda, Ida only has one option (the other side on the outside), and therefore so does Hannah (beside Gerda). There are two options of this type, giving us a total of $4 + 2 = 6$ possible seating arrangements on the bench.

Solution for Problem 19.32:

Since Clare learns three tones a day, of which she forgets one in the evening, she is actually learning two tones from one morning to the next. It therefore takes her 6 full days to learn $6 \times 2 = 12$ tones. On the seventh day, she learns three more for a total of $12 + 3 = 15$ in the evening.

Solution for Problem 19.33:

Any box that contains anything contains six boxes of the next smaller kind. Only the big box is not contained in some other box. This means that the total number of boxes is equal to $5 \times 6 + 1 = 31$. Since five of these contain other boxes, the number of boxes that do not contain other boxes is equal to $31 - 5 = 26$.

Solution for Problem 19.34:

There is obviously one big box. There are five of the next smaller size, and therefore $3 \times 5 = 15$ of the next smaller kind. Since each of these contains six tiny boxes, the total number of these equals $15 \times 6 = 90$. This means that the total number of boxes in the attic is equal to

$$1 + 5 + 15 + 90 = 111.$$

Solution for Problem 19.35:

Since there are exactly two pupils that like both groups, there are $9 - 2 = 7$ that only like Kyoto Motel and $7 - 2 = 5$ that only like Osaka Restaurant.

Since there are two that like both and 14 that don't like either one, the total number of pupils in the class is equal to

$$7 + 5 + 2 + 14 = 28.$$

Solution for Problem 19.36:
We can forget about the 11 that don't read any of the authors, so there are really only $39 - 11 = 28$ students to consider. Of these, we know that $6 + 5 + 1 = 12$ only read books by one specific author, so there are $28 - 12 = 16$ that read books by at least two authors. We can now define variables x, a, b and c to denote the number of students that read books according to the following table:

	A. Christie	E. Wallace	A.C. Doyle
x	yes	yes	yes
a	yes	yes	no
b	yes	no	yes
c	no	yes	yes

Comparing with the numbers we are given, we then get

$$16 = x + a + b + c,$$
$$10 = x + a,$$
$$9 = x + b,$$
$$11 = x + c.$$

Adding the last three equations gives us $30 = 3x + a + b + c$, and subtracting the first equation from this one then gives us $14 = 2x$, or $x = 7$. We see that there are seven students that read books by all three authors.

Solution for Problem 19.37:
It isn't too hard to write down all four possible options:

$$A - B - C, \quad A - C - B, \quad B - C - A, \quad C - A - B.$$

One way to see why there are exactly four is to consider what happens on the way to house C and back. Each of the other houses can be visited either on the way in or on the way out. This means that there are $2^2 = 4$ possible ways to deliver the mail without doubling back. Thinking in this way, we see that there would be $2^4 = 16$ ways to do it if there were 5 houses in the street, as he could deliver the mail to the first four houses either on the way in to, or on the way out from, the fifth house. Similarly, there are then $2^6 = 64$ ways if there are seven houses in the street.

Solution for Problem 19.38:

First of all, we note that there are four 13-year-olds playing in the tournament. Each of the four plays against three others, and since counting $4 \times 3 = 12$ would count every game twice (we are counting A playing B separately from B playing A), the total number of games between them is equal to $12 \div 2 = 6$. Similarly, there are six 14-year-olds playing, as each of the six plays five others, and we obtain $(6 \times 5) \div 2 = 15$ games among them. Since each of the four 13-year-olds plays each of the six 14-year-olds, we obtain a total number of $4 \times 6 = 24$ games of the required type.

Solution for Problem 19.39:

There are three routes (across bridges A, D and E) from the mainland onto the island and four routes (across bridges B, C, D and E) back to the mainland. Any of the three routes onto the island can be paired with any of the four routes back, giving us a total of $3 \times 4 = 12$ different routes leading from the mainland to the island and back.